Selling Safety

Lessons from a Former Front-Line Supervisor

T0332445

Selling Safety

Lessons from a Former Front-Line Supervisor

Patrick J. Karol

CRC Press
Taylor & Francis Group
Boca Raton London New York

CRC Press is an imprint of the
Taylor & Francis Group, an **informa** business

CRC Press
Taylor & Francis Group
6000 Broken Sound Parkway NW, Suite 300
Boca Raton, FL 33487-2742

International Standard Book Number-13: 978-0-367-42170-0 (Hardback)

This book contains information obtained from authentic and highly regarded sources. Reasonable efforts have been made to publish reliable data and information, but the author and publisher cannot assume responsibility for the validity of all materials or the consequences of their use. The authors and publishers have attempted to trace the copyright holders of all material reproduced in this publication and apologize to copyright holders if permission to publish in this form has not been obtained. If any copyright material has not been acknowledged, please write and let us know so we may rectify in any future reprint.

Trademark Notice: Product or corporate names may be trademarks or registered trademarks, and are used only for identification and explanation without intent to infringe.

Library of Congress Cataloging-in-Publication Data

Names: Karol, Patrick J. (Patrick John), 1959- author.
Title: Selling safety : lessons from a former front-line supervisor / by Patrick J. Karol.
Description: First edition. I Boca Raton, FL : CRC press/Taylor & Francis Group, 2020.
Identifiers: LCCN 2019040872 (print) I LCCN 2019040873 (ebook) I ISBN 9780367421700 (hardback) I ISBN 9780367822408 (ebook)
Subjects: LCSH: Safety education. I Industrial safety--Management. I Leadership. I Employee motivation.
Classification: LCC T55.2 .K37 2020 (print) I LCC T55.2 (ebook) I DDC 658.4/08--dc23
LC record available at https://lccn.loc.gov/2019040872
LC ebook record available at https://lccn.loc.gov/2019040873

**Visit the Taylor & Francis Web site at
http://www.taylorandfrancis.com**

**and the CRC Press Web site at
http://www.crcpress.com**

CONTENTS

PREFACE

I never had any aspirations to pursue a safety career. Not even the slightest intention. It more or less chose me; some might say that I was in the right place at the right time. I will share the details of that story later in the book. I can say without reservation, I am thankful I did land in the safety profession. Being a safety professional has given me a great sense of purpose and satisfaction in my professional and personal life. As a result, I have learned much since leaving my front-line supervisor position to become a corporate safety analyst in January 1994.

Much has changed since I was a supervisor almost three decades ago. Leadership and management theory has evolved, tools and technology have progressed at a pace few could have predicted, generations have changed, yet I see the same scenarios playing out today with safety professionals, supervisors, and operations managers. The same mistakes I made three decades ago are still being made today, which brings me to the purpose of this book. I have three distinct reasons for writing this book.

First, I want to share my story as it relates to Occupational Safety Management and the lessons I have learned about safety, starting with my first job out of high school and throughout my 25+ years as a safety professional. This is not a book about compliance with safety regulations or the Occupational Safety and Health Administration (OSHA). It is not about conducting a Job Hazard Analysis or incident investigation. It is not a book on the psychology of safety. There are already some very good books and resources available on all of these topics. It is the story about my experience with occupational safety as an hourly factory worker, front-line supervisor and safety professional. My story is about the mistakes I made in my first attempt at managing employee safety in my position as a front-line supervisor and the lessons I have learned since. Like so many other supervisors, I thought I was fixing a problem when I handed out disciplinary action for a safety infraction. It is about what I would do differently today if I was starting over.

I am sharing these stories because I believe the lessons learned will add value for operations managers, front-line supervisor or safety professional who has been befuddled by the "failed to follow procedures" issue. In other words "How do I get employees to do what I want them to do?" My hope is that I can reduce the learning curve for one person who has responsibility for employee safety. It is a learning curve that took me years to get over.

Second, I want to share my passion for occupational safety with you. More specifically, I want to share my passion for creating a work environment where all employees feel safe from occupational injuries and illnesses and feel empowered to grow. My hope is that by sharing my passion for safety, it will ignite a similar passion in you. If that passion already exists, then maybe I can help that passion burn a bit brighter.

Third, many very good people took the time to mentor me, and I will be forever grateful for the unselfish interest they took in me and my profession. As safety professionals, operations managers or human resources professionals, they provided different perspectives. In a small way, this book is my thanks to them. They came into my life sometimes only briefly, but always leaving their indelible mark on me. Their mark made me a better safety professional and person. I hope to thank my mentors by sharing those lessons learned with young safety professionals, operations managers and supervisors who struggle with the same safety management issues that I struggled with early in my career.

From a more practical perspective, my book seeks to fill a skills gap I see in many safety professionals, supervisors and operations managers: a gap between the technical skills and soft skills. I spent years developing my technical skills; after all, I was an operations supervisor who was trusted with a corporate safety role overnight. I had limited practical experience and even less technical knowledge. Learning technical skills first was the right thing to do. Having technical knowledge established my credibility and opened doors to conversations about hazards and risk control. What I failed to appreciate was the importance of the soft skills needed to influence change including communication, presentation and speaking skills, ability to influence without authority, and selling. I will consider this book a success if I can help guide the course of just one safety professional, front-line supervisor or operations manager and as a result, increase the chance of just one more employee going home safe.

ACKNOWLEDGMENTS

My appreciation for their assistance, guidance and input goes out to the following family members, colleagues and friends:

My wife, Nancy Karol for encouraging me and allowing me to spend the time it takes to write a book.

My son, SSgt Pete Karol, who taught me the importance of discipline and that if you are on time, then you are late.

My daughters, Nikki and Kristina, who remind me every day what can happen when you focus on your passion, and to have fun along the way. They have made me a better person.

Lanette and Marlon Wolcott for being so gracious and allowing me to spend time at their humble abode. For Marlon who was always so timely with a cup of coffee or other beverage depending on the time of day.

Najib Saadeh for his professional input and editing during several versions.

Sherry Pond for her input on several chapters, sitting through more than one of my presentations and for sharing stories over barley pops.

Lara Malatesta for providing her honest and professional feedback.

Carol Henson and the folks who organize the Georgia Safety, Health and Environmental conference ("The Savannah conference") for giving me an opportunity to speak every year. That is where it started.

Andrew Salvador for providing his insight on my presentation and several chapters.

Camille Oakes for attending my presentations and providing input on several chapters, and for being the best speaking partner ever.

Scott Ross for providing encouragement and input from an operator's perspective, and giving me my first speaking opportunity as an independent consultant.

Ken Armstrong for providing early guidance on the outline, for always answering the phone and being willing to share his knowledge and experience, but most of all for being a great friend and colleague since day one of my professional safety career.

Robin Blair for being my Spartan OCR running partner and helping me maintain my stamina.

Helen Morris for lending an understanding ear, counsel and encouragement when I needed it most.

"The Jims." Jim Swartz and Jim Stephan for giving me my first job in the safety profession. Their mentorship and example helped to light the fire of passion for safety in me. Jim Swartz for his career advice while I was unemployed.

In memory of my father Peter J. Karol, WWII veteran and 40 year employee and supervisor of that glass factory where I started my work career.

In memory of Don Rucker, who gave me my first safety responsibility to run the safety committee as a front-line supervisor.

ABOUT THE AUTHOR

 Patrick J. Karol is the President of Karol Safety Consulting, LLC, and holds a BBA from Georgia State University. Pat holds certifications as a "Certified Safety Professional, Associate in Risk Management, Safety Management Specialist and Certified Instructional Trainer."

Pat's work in the safety field began as a front-line supervisor with safety as a collateral duty, and now involves advising senior leaders on strategies to reduce risk. His experience includes over 20 years in the corporate safety departments of two Fortune 200 companies and the Federal government. Pat currently works as an independent safety and health consultant specializing in strategic safety planning and motivational speaking.

Pat's professional safety career includes:

- Sr Director of Safety with EEC Environmental.
- Safety Manager with ARAMARK Corporation Safety & Risk Control Department. Responsibilities included working with senior leadership to develop strategies and tactics to reduce risk of injury, environmental exposure and food borne illnesses.
- Safety and Health contractor to the Transportation Security Administration where he managed a team responsible for compliance at all US airports in the southeast region.

- Safety Manager with Delta Air Lines Corporate Safety Department. Responsibilities included investigation of serious accidents, development of procedures and training, identification of trends and development of mitigation strategies, and assisting with OSHA VPP applications.

Pat is a successful speaker and has spoken on safety and health at numerous regional and national conferences. He spoke most recently at Safety 2019, AIHce 2019, the Georgia Chapter of ASSP Professional Development Conference 2018 and the Georgia SHE Conference 2018. Pat is a past President of the American Society Safety Professionals (ASSP) Philadelphia chapter and was honored to receive the Gold Level award for 2014/15. He is currently the Area Director for the ASSP Keystone (Pennsylvania) Area for 2017–2021. He can be reached at www.karolsafety.com.

INTRODUCTION

I gave a presentation a few years ago titled *Selling Safety to Upper Management*. After all, that is what I did. I worked with organizational leaders to develop strategic safety plans and influence change. The feedback was generally good; however, more than one person responded that upper management was not the problem. The problem was with their employees. Specifically, "How do I get my employees to do what I want them to do?" "How do I sell safety to my front-line employees?" It is true that you need to sell safety up and down the organization to be successful. Selling safety to upper management is a bit different; in fact, it is quite different than selling safety to the front-line employee.

Safety should be like every other aspect of the business when it comes to how it is managed. Expectations based on a vision, mission and objectives should be clearly defined and communicated. A system of accountability should be in place, as well as key upstream and downstream metrics to determine if progress is being made. The problem is that we often try to manage safety differently. We use a command and control management style that focuses on compliance with governmental regulations. Investigations focus on the employee's behavior rather than the system or process. When the root cause is identified as "failed to follow procedures," we use disciplinary action to punish the transgressor. Then we retrain everyone, which is often viewed as a form of punishment. Games and giveaways are used as a reward or simply to keep their attention as if we are afraid they might wander off. Banners with safety slogans are hung throughout the work area where they eventually collect dust and become obsolete.

I know because that is how I managed safety as a supervisor. That was my approach because it was how I was taught, and besides, everyone else took the same approach. After all, I am the supervisor so "it can't be my fault." If it is not my fault it must be the fault of that poor employee that was involved in the incident. I was only following the corporate Standard Operating Procedures (SOP) for incidents that called for progressive disciplinary action.

At least it was clear, but also clearly wrong, as I would find out years down the road. It is little wonder managers and safety professionals alike have difficulty getting employees to do what we want them to do. Much has changed since I was a supervisor almost three decades ago, but surprisingly, I still see some of these same tactics used today. These tactics encourage compliance, but stifle creativity and initiative. "I'm just doing my job," rather than "I see a hazard and I'm going to correct it."

The problem exists today in varying degrees. How do we get employees to do what we want them to do? How do we convince them that safety is important and procedures should be followed? How do we exceed minimum standards? We cannot and should not codify every situation or task. If we did, every job or task would have a novel size set of procedures, which discourages people from taking responsibility and using their best judgment. The challenge then becomes how we get employees to do what we want them to do: follow procedures, recognize hazards and take action. I have faced these challenges throughout my work career. Fortunately, we have several tools in our toolbox that can help.

This book is based on my experience with safety as an hourly employee, front-line supervisor who wore safety as one of many hats, a full-time safety professional for two Fortune 200 companies and the federal government and now as a safety consultant.

There were three pivotal events in my working career that shaped my approach to safety and the tools I use. I will share the story of those events with you because it provides some context to the lessons I learned. The lessons learned are based primarily on my experience, but also on teachings from current-day thought leaders inside and outside the safety field.

Selling Safety to the front-line is about influencing employees to do what we want them to do; go beyond compliance by taking responsibility for identifying and correcting hazards before they become an incident.

THE PIVOTAL MOMENTS

1

THE FIRST PIVOTAL MOMENT:
I TOLERATED SAFETY

If I have to wear these things, you have to!

– My foreman to me when he caught me
not wearing my safety glasses

The first pivotal moment, which was my first exposure to occupational safety and health, came during my first job after high school graduation in 1977. I worked at a glass bottle factory in Hapeville, Georgia, on the south side of Atlanta. I went to work there because that is where my dad worked as well as so many other members of my family, so it seemed natural to follow suit. Besides, I did not feel like college material.

At the time, most containers were made of glass: milk jugs, Gatorade bottles, prescription bottles. After the bottles went through the cooling process, decorating and packaging, my job was to take the boxes off the conveyor and stack them on a pallet.

By the time the bottles reached me, they still radiated heat from the forming process. The conveyor moved at a steady pace and did not stop for breaks. My motion was constant. It was not a frenzied pace, but it was steady and it was hot, always hot. I also remember the factory being noisy, although I do not recall wearing hearing protection. Broken glass on the floor was a constant as were people assigned to sweep up the broken glass.

I do not recall any specific safety training as part of my orientation. I also do not recall being placed under the tutelage of a senior employee. I suppose the knowledge required to take a box of bottles off a conveyor and place them on a pallet did not take a great deal of training. The plant had a requirement to wear eye protection while on the factory floor due to flying glass chips. Work shoes with thick soles were necessary. The eye protection was supplied by the company seemingly without consideration to fit. Further, the plant was hot and the glasses tended to slide down my nose. The problem was compounded the more I sweated. So, I did what any good junior employee would do in this situation – I learned from the senior employees. They clinched the stem of the glasses between their teeth and quickly put them on when they saw the inspector or foreman on the floor. I soon mastered that maneuver. That worked great until one day, the foreman walked up from behind me. The foreman had a crew cut and stood over 6 ft. tall. He reminded me of a drill sergeant. He tapped me on the shoulder and said in no uncertain terms, "If I have to wear these things, you have to wear them. Put 'em on!" His words and tone still ring out in my mind today. That was the first pivotal moment that shaped my perception of safety. It was not a positive experience.

Safety was something I had to do. I learned to tolerate safety at best. It was something I endured. It was boring and aggravating. I did not appreciate or even see the benefits; it was enforced because we had to, not because we wanted to. Safety was and still is often thought of as a negative. Terms like "accidents," "investigations," "disciplinary action," "citations" and "penalties" all conjure up negative thoughts and images. There are different methods for selling safety, and that was one ineffective method, although it ensured we met minimum standards at least most of the time.

Safety continued to be something I looked out for, not because it protected my well-being, but because I knew I would feel the wrath of the foreman if I was caught being non-compliant. That perception of safety stayed with me for several years. Unfortunately, there was so much around me that reinforced my negative image of safety. Other issues were seemingly more important, like the relief guy showing up on time so I could take my break, and if my paycheck accurately reflected my overtime. Safety just did not rank as anything important to anyone, including me.

2

SOAPBOX SAFETY

That CANNOT happen again or you WILL get written up!

– Me, as a supervisor to my team after an incident.

I worked long enough at the glass factory to figure out that a college degree just might serve me well. I headed to college and studied business administration. Before graduating, I landed a summer job in the airline industry at the Atlanta airport. I was excited!

The airport ramp environment was a hectic cluster of aircraft, motorized vehicles of all sorts, rolling stock (used for hauling cargo, airmail and baggage) and people. There were baggage tugs, jet tugs, belt loaders (motorized conveyors), fuel trucks, lavatory trucks, catering trucks, maintenance lift trucks and water trucks, all designed to service an aircraft quickly. Pedestrians constantly intermingled with vehicles – sort of a concrete, motorized version of the wild, wild west. Break areas were like the saloons. They emptied out directly onto the ramp where your horse (bag tug) was tied up.

Baggage, cargo and mail had to get loaded on the correct flight at the right time. Flights arrived in "pushes"; in other words, they were not spread out evenly over the day. They arrived and departed in banks of flights similar to traffic jams in the morning and afternoon. Lockheed L1011 Tri-Star, Boeing 727, McDonald Douglas DC-9 and DC-8 aircraft filled the ramp. At the time, there were no traffic lanes to streamline vehicle traffic flow, no red lights to direct vehicle traffic and no safety officers to direct aircraft, vehicles or pedestrians. Only the people in the flight control tower were directing aircraft. As a pedestrian on the ramp, you were on your own, sort of a buyer beware situation since aircraft pilots do not have a clear view of the ground, and even if they did, at around 114,000 kg (250,000 lbs), aircraft cannot exactly stop on a dime. With engines that can produce 180,000 kg (400,000 lbs) of thrust, creating suction at the engine intake (front of the engine), the ramp environment was a dangerous place to work.

My job was to load baggage, cargo and airmail in a timely manner. I received about one week of classroom training before stepping out on the airport ramp. I still vividly recall my training on the hazards of Aircraft Engine Blast and Suction. You cannot see it and only barely hear it (we wore hearing protection), but it was ever-present. It was a silent killer. Stories of severe accidents resulting from blast and suction stayed with me. With so much aircraft and vehicle noise, all sources of noise became just white noise. Unless you saw trash or dirt being kicked up by jet blast, you may not notice that an engine was running. We were taught to look for the red blinking lights above and below the fuselage, which indicated the engines were running. The lights indicated the presence of jet blast and suction hazards, but that was just one more thing to look for.

I was paired with a "buddy" and told to "stick close to Rick" for a couple of weeks. Rick was a senior guy with an interest in mentoring (and patience). It was Rick's job to show me the ropes, assess my performance and let supervision know when I was ready to stand on my own in this airport version of the wild, wild west.

As part of my informal "buddy" training, Rick escorted me out to the middle of an aircraft taxiway and pointed to a specific spot and said, "Stand here." Rick did something that morning that has stayed with me to this day. He told me the story of a horrific accident involving another "summer help" worker who was run over by an aircraft. My "buddy" Rick taught me many lessons over the course of the next few days, but nothing stuck with me like that message. His message stuck with me because it was conveyed in a story. It was a story that I easily identified with as a new employee and teenager.

My job orientation did not stop there. Rick showed me the perspective from the ground and from the flight deck of an aircraft. He emphasized the obstructed view from the pilot's perspective and the noise on the ramp (aircraft engines easily exceeded 90 dBs at the time) that hindered my ability to focus. The noise essentially took away one of my senses, effectively reducing my ability to hear and thus anticipate changes that may present new hazards. His message was simple: be on the alert for these hazards (moving aircraft). Recognize that your environment (high noise, weather conditions, congestion) will change and hamper your ability to recognize hazards. I eventually went on to work around aircraft for the next 13 years with a very healthy respect for the environment and hazards.

After my brief stint as a "summer help," I worked several hourly jobs including janitor (I cleaned bathrooms in the aircraft hangars), forklift operator, aircraft cabin cleaner, lavatory truck driver (yes, it is what it sounds like) and baggage handler where my job was to load and unload bags and cargo from aircraft. I eventually worked my way up to ramp area supervisor. I was responsible for managing the workload, communicating plans and personnel scheduling for an area that covered about 4–5 gates and about 30–45 employees.

Every company has a culture often defined by at least one or two metrics that drive action. In the airline industry, after passenger safety it was on-time performance. The US Department of Transportation (DOT) published results every month in the form of a ranking by airline. These results drove a real sense of urgency based on your rank. On-time performance drove our culture; it is what management paid attention to and talked about, and often yelled about. If you were responsible for a flight delay, you had to explain verbally and in writing up the ladder and hope they accepted your explanation. Departments argued over who was responsible for a delay. It created a sub-culture of blame similar to the safety culture. "It can't be my fault, so who can I blame?"

Our safety culture at the time could be summed up in one phrase: "whatever it takes to get the job done!" It translated to mean "whatever it takes to the get flight out on time." If someone suffered an occupational injury, well that was just the cost of doing business. The protocol was to send the injured employee for medical attention (we had a nurse on-site) and home until fully recovered. Light or restricted duty was not offered at the time.

Aircraft damage prevention was the primary focus of the safety initiative. Repairing an aircraft struck by a motorized vehicle was expensive. In addition to direct costs associated with the repair, there were delay costs. A delayed flight had a domino effect. One flight delayed impacted subsequent flights later in the day. Passengers had to be rebooked or overnighted if another flight was not possible. The brand also took a hit. Frequent flier surveys always listed on-time arrival as very important, not to mention the safety implications of flying a damaged aircraft. If an aircraft was damaged and it resulted in a delayed flight, and it almost always did, all hell rained down from above. I called it "Soapbox Safety." At the supervisor's shift briefing the next day, the boss mounted the soapbox and preached, berated, cussed and fumed about our lack of diligence and care working around aircraft. As a young supervisor, I followed suit. I went back to my area, got on my soapbox and preached and yelled and cussed. It started somewhere around the top and worked its way down. I learned how to manage safety from their example (Figure 2.1).

Figure 2.1 Soapbox Safety. Illustration by Masha Balac.

We had two "safety" tools in our arsenal: "soapbox safety" was the first and the second was disciplinary action. It was progressive and worked something like this:

1. First offense – a letter of reprimand is placed in the employee's file
2. Second offense – one day off and a letter of reprimand is placed in the employee's file
3. Third offense – three days off and possible demotion or termination.

Offenses typically meant aircraft damages. Employees were not disciplined for incurring an occupational injury. After all, it was a tough, physical environment and injuries were likely, if not expected, to occur. As a supervisor, if I had a team member out of work due to injury, I covered the workload by paying overtime. I became somewhat of a hero for the moment when I could offer overtime to someone looking to make a few extra bucks. Word would get around pretty quick, "Hey, PK has a couple of folks out on injury and is paying OT this week." It is what I call a disincentive to injury prevention.

But that one-two punch is essentially how I managed safety as a front line supervisor on the ramp. Yell a lot, cuss some, hand out disciplinary action when necessary. What was amazing about this one-two punch was that it worked … at least for a few weeks. Sooner or later the whole scenario was repeated.

To sum it up, on-time performance was the key metric. Aircraft damage impacted on-time performance; therefore, the prevention of aircraft damage was managed aggressively and drove the safety culture. Arbitrating over a safety incident and sentencing employees to disciplinary action gave the illusion that my management style was working. In reality, that punishment was only improving performance to minimum standard levels and leaving the employee with a negative experience. I talk more on safety as positive in Chapter 8.

3

THE SECOND PIVOTAL MOMENT: SAFETY BECOMES PERSONAL TO ME

I heard the call come over my radio, *Echo 1 to gate A-17...*

The second pivotal moment came more than 10 years later. I was a supervisor working the evening shift, 3:30 pm to midnight. About 12 supervisors were on duty spread out over the Atlanta airport, an airport that at the time spanned across six concourses, four runways (today five) and one terminal (today two). Then, it was one of the busiest airports in the world and covered over 4,000 acres. Since 1998, it has been the biggest and busiest airport in the world in terms of flights and passengers and covers 4,700 acres, according to their website.

All supervisors carried radios for the purpose of communicating with each other and other departments. We could speak directly to the ramp tower operation, agents at the passenger service gates, baggage service and other supervisors. Radios were also invaluable in emergency situations.

It was dusk one cool, fall evening in 1992 when I heard a call coming over my radio, "Echo 1 to Gate A-17..." I could sense some urgency in the call and recognized the voice; it was "John" my counterpart in that area. Echo 1 was the ambulance service dedicated to the airport, and a call to Echo 1 usually meant a medical emergency. Gate A-17 was directly across the ramp from my area. I glanced up and could see that an aircraft had started to push back from the gate and stopped after a few feet.

It was not uncommon for the ambulance service to be called for passengers who were ill. If a passenger was removed from a flight, it was our job to remove any checked bags. That involved obtaining a description of the bag, maybe the passenger's name or bag tag number and then manually digging through as many as 300 bags to find the passenger's bag. So it was my assumption that a passenger was sick and needed to be deplaned. I was between pushes in my area and thought I would ride over to Gate A-17 and offer to help "John." I knew him well and knew he would do the same for me.

It took me less than a minute to find a tug and drive to Gate A-17. As I arrived, I noticed a couple of guys had gathered around the left main gear of the aircraft. These are the quad tires and double axel under the wing. I slowed down as I approached and the thought occurred to me, "We didn't do what I think we just did." My fears were confirmed when I noticed "John" taking off his coat and putting it on an employee who was lying down. Another coworker, "Steve" the fueler and another good friend of mine, had removed his belt and was tying it around the employee's leg. We had just run over "Vance" with a passenger aircraft, an aircraft that weighs 114,000 kg (250,000 lbs) empty.

I looked up to see if I could make sense of what had just happened when I noticed the jet tug operator. He was slouched over the steering wheel with his face hidden in his folded arms. He realized what had happened. At that moment, the ambulance arrived. I knew there was nothing I could do. I never even got off my tug. I simply turned around and drove back to my area. I would have another push arriving soon. I told myself, "I gotta get back to work." I did not know what else to do. It was an incredibly helpless feeling. I kept mumbling to myself, "How did that happen?" It was at that pivotal moment that safety became personal to me. I knew I could no longer just "tolerate" safety.

"Vance" lived, but lost one leg above his knee. "Vance" was obviously the primary victim who suffered both physically and emotionally. There were also several "secondary" victims affected. The jet tug operator was a secondary victim as well as "Vance's" wife and two young children. I can only imagine how an incident like this affects someone who is directly involved with or may have contributed to the incident.

Although at the time I did not give it any thought, "John" the supervisor and "Steve" the fueler were also secondary victims. Today, I wonder how they were affected by the incident. I am sure as with any supervisor or manager whose team member suffers a serious injury on their watch, "John" no doubt second-guessed himself and wondered what he could have done differently to prevent this incident. In the following weeks, I often wondered what I could do as a supervisor to prevent a similar incident.

If this incident affected "John" on the job, it certainly affected him at home. What about "Steve" the fueler? How did this incident impact him? His job was to drive a fueling truck on the airport ramp around aircraft, hundreds of motorized vehicles and people. Did this incident distract him in the days following the event? Did this incident affect the pilots of the aircraft? How did it affect "Vance's" coworkers? Did they do anything different to reduce their risk of incurring a similar incident? I know how it affected me and I was not directly involved. I can only imagine how it affected others who were more directly involved or who knew "Vance" well.

I later learned more detail around the incident. Each aircraft push-back or movement required "wing-walkers." It is just as it sounds. A person is positioned at each wing to ensure clearance between wingtips and ground equipment, while another person pushes back or moves the aircraft using a jet tug. On this night, the flight was delayed several times. Underneath the fuselage behind the main gear tires on a B-757 is the auxiliary power unit (APU), essentially a small engine that provides power to the aircraft when the main engines are not running. The APU blows warm exhaust toward the ground. Since it was getting colder, the wing-walkers, who had been standing on the ramp, decided to walk underneath the aircraft to warm up next to the exhaust, a somewhat common occurrence.

The call finally came from the control tower to the jet tug operator for clearance to depart. I believe due to the extensive delay and strong emphasis on on-time departures, the jet tug operator started pushing the aircraft back immediately not realizing the wing-walkers had left their position. As "Vance" realized the aircraft was moving, he ran toward his left wing-walker position. Unfortunately, the main aircraft gear ran over his leg as he ran to his position.

As is so often the case, safety culture change is initiated after a serious incident and this was no exception. Organizations take a hard look at themselves and are dissatisfied with what they see. Over the course of the next few months, management benchmarked other airlines and industries ultimately deciding they needed a dedicated ground safety department. Up until then, safety was piecemealed out to either training, legal or the personnel department. A safety department was formed, and a director was hired from outside the organization, something unheard of prior to this event. I was fortunate to be chosen to join the ground safety department in January 1994. That was the beginning of my professional safety career.

I worked in the department for the next 13 years learning from my director and manager who came both from within the aviation industry. I learned to appreciate safety when I could finally see the impact of my projects on the front-line employees. We changed a culture from "whatever it takes to get the job done" to a culture that valued employee safety and understood the job could still get done while operating safely. Unfortunately, it took a serious incident.

The change did not happen overnight. We had a mandate from the top, but it still took several years to reach all levels of the organization and make a lasting change. We did it by establishing upstream and downstream safety metrics, developing new procedures and training that addressed known hazards, investigating incidents to identify contributing factors and tracking corrective actions. We implemented engineering changes and benchmarked with our competitors. We identified key allies (we called them Safety Coordinators) in the field and engaged them in the process. Safety was now personal to me and I worked hard to develop and update tools, investigate incidents and implement a safety management system. I gave many safety presentations to management at regional meetings and never missed an opportunity to talk safety.

I left the airline industry after 26 years, half as a front-line employee and supervisor and half in the corporate safety department, but I carried many lessons learned that I continue to apply today.

4

THE THIRD PIVOTAL MOMENT: I HAD TO MAKE SAFETY PERSONAL TO OTHERS

That can't be good news.

        – Me, to myself when the phone rang at 6:30 am.

The third and final pivotal moment came almost 15 years later. After a brief stint as a safety and health contractor to the federal government, I landed with a food and facilities company in Philadelphia, a global company with over 200,000 employees worldwide.

I am not sure who the call came from on that late summer morning in 2013. I had just arrived at the Philadelphia airport for a trip to a city that I don't recall. I do, however, distinctly remember *the* call, which was very brief. Anyone who has been in safety long enough knows that good calls don't come at 6:30 in the morning. This was no exception. The caller relayed a message that anyone in the safety profession hopes to never receive. An employee was struck by a vehicle at work and killed. I switched destinations and arrived on-site later that morning.

I did not know "Sandy," but as I walked into the break area, I noticed the memorial created by her coworkers and a flood of memories from the first pivotal moment years earlier returned. It was quiet even though the break area was full of employees. Her coworkers were sitting around the breakroom table trying to console each other and make sense of what just happened. Her coworkers were male and female, Hispanic, African-American and Caucasian, but they all felt the same pain and disbelief that morning. "Sandy" was a 22-year-old mother who would never see her two-year-old daughter grow up.

Somber does not begin to describe the mood or atmosphere at the worksite. It was as solemn and mournful as anything I have ever experienced. There was no lively conversation or hustle and bustle of hurrying to eat lunch or finish a cup of coffee. It was just quiet. Safety was personal to me and I could see that it was personal to her coworkers now, if it was not before. It was at that pivotal moment I realized, I need to make safety personal to everyone else.

Yet, I had a job to do and I needed to interview employees as part of the investigation. I did not want to do it any more than they wanted to talk about it. I knew that a hallmark of strong safety cultures was reacting with a sense of urgency to all incidents, especially serious incidents. I kept telling myself that successful companies view these incidents as an opportunity to improve, and to improve, we needed to find out what happened. Something in our process broke down, and it was my job to uncover the details. We owed at least that much to "Sandy" and her coworkers.

I later learned that the incident occurred outside during the predawn hours in a dimly lit area with no aisle markings, pedestrian markings or vehicle markings. The employee was not wearing reflective clothing, nor was it required. "Sandy" was part of a team that was contracted to clean and stock the aircraft cabin. The task was time-sensitive with a need to be completed later that morning, but something was different that particular morning, or at least out of the ordinary. That morning, mobile stairs were not available to gain access to the aircraft cabin. The crew of about five waited on the airport tarmac next to the aircraft for someone to arrive and provide access.

An employee of the client was eventually contacted to gain access to the aircraft cabin. There were reports of being shorthanded. The driver of the vehicle, a utility pickup truck, was in a hurry and never saw her in the dimly lit area.

As is typical, there were multiple contributing factors and always, something different. On this early morning, the team was unable to gain access to the aircraft cabin and had to wait in a poorly lit area in the predawn hours. Something that rarely occurred.

Similar to the ramp incident so many years earlier, my thoughts were not with the driver. The reality is that a person suffers the guilt of having been directly involved in a workplace fatality. The driver was a secondary victim as was "Sandy's" daughter, other family members and coworkers who witnessed the incident or heard about it.

Before I completed the investigation, the thought occurred to me, "How do I get the message of this incident out to over 200,000 employees worldwide?" How do I make safety personal for so many people over such a large geographic area? How can I make safety personal to every front-line employee, supervisor and middle manager? I cannot write enough procedures, conduct enough training or script enough messages for senior leaders to get even one message to every employee in a timely manner. It takes one incredible network of people who feel the same about safety. It takes a network of people who are passionate about safety and willing to step forward and take action. Simply sending a message out and hoping that message is cascaded down to every employee without being misinterpreted is not a recipe for success. It would take a network of missionaries. More on that in Chapter 8.

There were other impactful moments in between, but none had a profound impact on me like these three events. None lit the fire of my passion for creating a safety culture where employees could do what they do best without the worry of being injured.

I share these pivotal moments in my working career with you because each shaped my approach to safety and how I believe it should be managed. These moments are my experiences that provide the context to the following chapters where I describe how to sell safety to a specific group of workers, the front-line employees.

5

WHY SELL SAFETY?

The three most important things about safety I learned early in my professional safety career are 1. Selling 2. Selling 3. Selling.

– Patrick J. Karol

Two components are key to the success of a safety profession: the technical component and the soft skills component. You might call it the science and art of safety. Both are essential and both are different. The technical component is our ability to understand which chemicals react together to form a deadly liquid or gas. It is our ability to understand and calculate noise reduction ratings and assess hazards associated with in-running nip points. It is the knowledge to differentiate between upper and lower explosive limits. The technical component establishes our credibility. It opens the door to conversations about hazards, risk and risk reduction. We cannot succeed without first having a level of technical skills and knowledge. That holds true for supervisors, managers and safety professionals.

While the technical skills and knowledge are important, even expected, it is not enough. It is not enough if you want to influence change that results in a strong safety culture. You cannot get there by being singularly proficient in the science of safety. You must be proficient in both the art and science of safety. Let me be clear, I am not saying be a salesperson. I am saying, be a subject matter expert in safety and health who knows how to market and sell.

Still, why should we need to sell safety? It sells itself, right? Employees at every level should understand that safety is important and following procedures is common sense, right? Some people see the value and do not need to be influenced. Unfortunately, safety often comes with a poor reputation largely based on how it is managed. Often, that means managing by catching someone doing something wrong. We play the role of safety cop or safety officer by managing to compliance not because it prevents incidents, but because it is a regulatory requirement. Many employees have experienced safety as

I did as a young factory worker. As a result, we are left with a negative perception of safety. Having technical skills and knowledge will not change that perception. What we need is an understanding of the art of safety. We must know basic marketing, be able to present and sell. The point is, we can be a subject-matter expert only and ensure compliance or we can learn the art of safety and influence change at organizational and cultural levels.

We have established that we need two sets of skills, but the question remains, how do we get employees to do what we want them to do? We can have all the prerequisite technical skills, education, experience and knowledge of all standards, but if we cannot influence a worker to follow procedures, we will struggle to get beyond a minimum standard.

Let's examine how we often do things today. When things go wrong and an incident occurs, our first reaction is often to assume someone did not do what they were supposed to do. Our corrective action is to add more procedures, more people, re-training (because it obviously did not take the first time), more, more, more. Unfortunately, more is not always better. If the incident was particularly egregious, we look for someone to pin the blame on. As a young supervisor, I knew it could not be my fault: "I am the supervisor, so it must be somebody else's fault." We identify the culprit or culprits and sentence them to disciplinary action. A warning letter they sign admitting guilt and promising not to do it again goes into their permanent personnel file. Or maybe we pull out that document they signed three years ago as proof that we trained them (another popular tactic to divert blame). Maybe we recommend a demotion or even termination.

Two things happen next as we continue to spiral downhill. First, we do not fix the problem and second, any trust the "guilty" party had in us is gone out the window and good luck trying to get that back. In addition, after the sentence has been carried out, you can count on the guilty party to do exactly what is expected of them...and nothing more! The discretionary effort that tends to be so valuable yet so elusive is now out of the question. Their involvement or engagement in the safety process is a pipe dream. Yet our expectations are that the problem that caused the incident is corrected and the guilty party has been rehabilitated. Unfortunately, recidivism tends to run high in this situation. In addition, it is not likely that others learned from the incident.

Management by compliance is another common approach. It is transactional management or Management by Exception (MBE).

We catch them doing something wrong because it is easy to do. Time for the annual mandatory training. That is sure to fire up enthusiasm in a front-line employee. It is almost as if we blame the regulators for the mandatory training. To exacerbate the problem, we take training off the shelf or use the same training we have used for the past 10 years. It is obviously dated or, even worse, it does not apply. I have seen training programs being used that were over 20 years old because the trainer really liked them. The real problem is that it sends a message that safety is not worth updating or making relevant, further reinforcing negative perceptions of safety.

There is a way forward. All progress begins by thinking differently. In this case, thinking differently about how we manage safety: the approach we take and how we talk about safety, and ultimately, how we get employees to do what we want them to do.

Our two most important responsibilities are to ensure compliance with government regulations and to ensure everyone goes home safe. They are not mutually exclusive, but neither are they in perfect sync with each other. If we are attempting to fulfill both responsibilities by conducting audits, inspections and investigations, we are meeting minimum standards, but not bringing full value to our employees or organization. We are not achieving our full potential or impact as safety professionals, operations managers or supervisors. We are not contributing maximum value. It is not about doing more. It is about influencing others, inspiring others to want to do the right thing. We cannot do that by conducting more inspections or audits, and certainly not by handing out more disciplinary action.

We should ask ourselves, "What impact did I have today?" If the answer is, "I ensured we were in compliance with energy isolation procedures," then great! That is significant! You did a nice job, but you didn't bring maximum value. Another question to ask ourselves is, "Did I inspire or intimidate?" If you hesitated in answering this question, it might be time to rethink your approach. Selling safety is about influencing and inspiring others to want to do the right thing. It is about wanting to get involved for the sake of a safer community.

Safety is about making the working life of an employee better, which goes beyond regulatory compliance. When we do that, we help ensure every employee can enjoy the things they hold dearest, the things closest to them. Being ambitious when it comes to protecting the well-being of an employee means connecting safety to what is most important to them. That is the essence of selling. When we make that connection, we have created value by helping them understand the value that a safe operation brings to them personally. Essentially, when we sell safety we are creating value for the front-line employee, which in turn creates value for the company and all stakeholders. For the safety professional, doing these things leads to a fulfilling career.

Some safety professionals, managers and supervisors complain because they do not have the resources they feel they need to get the job done. Others create value without the need to expend more resources, but it takes some imagination, creativity and ability to sell. More specifically, it takes the ability to sell safety to front-line employees.

Selling is about influencing someone to do something you want them to do and for the right reasons. It is about getting front-line employees to follow safety procedures, to go beyond compliance, to report hazards, to identify near-misses, to get involved in safety committee work and to look out for their coworkers. Assuming that safety should just sell itself shirks our responsibility for creating safe work environments. There are a couple of challenges with this approach. Besides, it just does not work, unless you are satisfied with minimum standards.

First, following safety procedures often involves behavior change or establishing new habits, something that does not happen overnight or after a single training session. To establish new behaviors, the employee, or anyone for that matter, must be influenced to make those changes. If selling is about influencing the other person to do something you want them to do, then selling safety is about influencing an employee to follow safety procedures, to use training techniques, to report hazards, to join the safety committee and to go beyond simple compliance. Secondly, why would we or anyone take action (buying a TV or a car, for example) without knowing or understanding the value? The answer is, we would not. Safety is no different. We know how difficult it is to implement a new procedure. When we do not explain the reasons and value the new procedure brings, getting cooperation from people at various levels of the organization is difficult. That is where our ability to sell safety comes into play.

Make no mistake, safety is a tough sell. We want to see immediate benefits from our action. We want something for our effort, but safety does not work like that. If we do everything right, nothing bad happens or we can do everything wrong and get lucky, and nothing bad happens. No injuries, no damages. Everyone goes home the way they came to work. There is no immediate feedback, no obvious return on the investment of time and effort. To compound the challenge, you cannot measure an injury that does not happen. In either case, there is no immediate benefit, which in turn makes safety an even tougher sell.

How effective we are at selling safety is critical if we expect to gain cooperation from employees or if we expect them to expend their discretionary effort on safety. As the seller, we are a key component in whether or not an employee will choose to buy from us or to follow safety procedures. Customer loyalty is often defined by repeat business which is impacted by our ability to effectively sell safety.

Consequently, front-line employees buy safety from us based on how they feel about us just like we buy cars, TVs or electronics. They buy safety from us, not organizational leaders. That is not to downplay the importance of organizational leaders, but the safety managers, operations managers, and more often the supervisors are the face of safety. If safety is portrayed as something we have to do, then the message will be clear and result in a negative perception of safety.

It is also not about our expertly written procedures or training. It is about us and how we position safety as a value. In other words, it is how we sell safety.

Games, gimmicks and giveaways do not work either. Would the finance department use a game of bingo to determine your safety budget? Of course not and we should not either. I am not saying safety should not be fun or we should not celebrate our successes. A game of chance can be fun, but do you want to base even a portion of your safety program on a game of chance? It seems to send a mixed message. Avoid, reduce and mitigate risk; do not take chances with your safety, but a game of chance is OK? A game where there is one winner means there will be lots of losers. Even the Occupational Safety and Health Administration (OSHA, 2016) frowns on programs based on injury counts and for a good reason. They tend to influence the wrong behavior. They discourage reporting. Safety, after all, is about mitigating risk or chance, the chance of injury or illness. We must influence workers by showing them that following procedures and training techniques, safety can help them get what they want. Workers are willing to buy from us, but we need to be good sellers.

REFERENCE

"Final Rule Issued to Improve Tracking of Workplace Injuries and Illnesses - OSHA Interpretation, 1904.35(b)(1)(i) and (iv)," 2016.

6

HOW TO SELL SAFETY

The most successful leaders know where they are going, how to get there and believe in their heart it is the right thing to do.

– Patrick J. Karol

We can take lessons from the selling profession. The same strategies and tactics used to sell a product or service can be used to sell safety to a front-line employee.

Selling safety starts with knowing a little about being a good leader. We like to follow others we consider to be good leaders. Being a good safety salesperson starts with trust, basic leadership characteristic. The more workers trust us, the more receptive they will be of our message. Gaining trust starts with demonstrating our expertise. If a worker knows we have a level of knowledge and insight, they will see us as experts. They are more likely to listen to and accept messages and recommendations from experts. One way to position ourselves as experts is to share our insights and experiences related to the job and the job hazards. We associate experience with expertise, and sharing our insights provides value. For example, if you have workers who routinely use ladders, you may provide insight into fall injuries. Falls from portable ladders including step, straight, combination and extension ladders are one of the leading causes of death and injuries on the job. According to the Bureau of Labor Statistics (2018), "Over a six-year period from 2011–2106, there were 3,723 fatal falls to a lower level. The most common sources in these cases were ladders (836 fatal injuries)."

Include personal stories to show not just your level of expertise, but your level of caring and emotion. Stories that are emotional work because emotions are contagious (I talk more about stories in Chapter 7). It is like being a player-coach working on the floor with employees to get the job done. As a coach, we explain the purpose for procedures and as a player, we set the example by following procedures. In addition, we seek their input. We give them a voice in the process.

Of course, you cannot fake expertise. You need to do the research, study, review injury trends, read the government standards and letters of interpretation. Talk with colleagues, benchmark with other organizations and safety professionals and achieve certifications. When you have the facts and the context around the facts, you can speak with confidence. If you display confidence, workers will have confidence in you making them more likely to trust you. When they trust you to have the right answers or to find the right answer, they will follow you. Your recommendations will have the weight of authority.

Successful leaders are also successful influencers. They possess characteristics we admire and make us want to follow them.

They are empathetic. You can gain credibility and trust by walking a mile in another person's shoes. Have you shadowed an employee, trained to do their job and worked alongside an employee? We are at our best when we play the role of player-coach.

They are trustworthy. People gravitate toward those who are genuine because they know they can trust them. It is difficult to like someone when you do not know who they really are and how they really feel. Genuine people mean what they say, and if they make a commitment, they keep it. You will never hear a truly genuine person say, "Oh, I just said that to make the meeting end sooner." You know that if they say something, it is because they believe it to be true. If you believe "zero" injuries or defects is achievable, do not back down, do not compromise.

They are genuine. Genuine people are open-minded, which makes them approachable and interesting to others. It is difficult to have a conversation with someone who has already formed an opinion and is not willing to listen. Having an open mind is crucial in the workplace, as approachability means access to new ideas and help. To eliminate preconceived notions and judgment, you need to see the world through the employee's eyes. This does not require you to believe what they believe or condone their behavior; it simply means you quit passing judgment long enough to truly understand what made them do what they did. If they failed to follow a procedure, there is a reason. Having an open mind means finding out why, and "can't fix stupid" is not a reason.

They are humble. They value everyone's opinion. They understand success comes from having the right people around them and listening. They know they need others because they do not have all the answers or even have the best answers. When conducting an investigation, I learned that I did not need to solve the problem myself. I only needed to ask an employee. More times than not, they knew the problem and how to correct it.

They do not boast. We have all worked with people who cannot stop talking about themselves and their accomplishments. Have you ever wondered why? They boast and brag because they are insecure and worried that if they don't point out their accomplishments, no one will notice. Genuine people do not need to brag. They're confident in their accomplishments, but they also realize that when you truly do something that matters, it stands on its own merits, regardless of how many people notice or appreciate it.

They are generous. We have all worked with people who constantly hold something back, whether it is knowledge or resources. Genuine people are unfailingly generous with whom they know, what they know and the resources they have access to. They want you to do well more than anything else because they are team players and they are confident enough to never worry that your success might make them look bad. In fact, they believe your success *is* their success.

They treat EVERYONE with respect. Whether interacting with a client, front-line employee or servers taking their lunch order, genuine people are unfailingly polite and respectful. They understand that no matter how nice they are to the people they have lunch with, it is all for naught if those people witness them being rude toward others. Genuine people treat everyone with respect because they believe they are no better than anyone else. This can be a challenge for anyone with safety responsibilities, especially when an incident occurs. We see this as a reflection on our performance and automatically look for someone or something to blame.

They are thick-skinned. Genuine people have a strong enough sense of self that they do not go around seeing an offense that is not there. If somebody criticizes one of their ideas, they do not treat this as a personal attack. There is no need for them to jump to conclusions, feel insulted and start plotting their revenge. They are able to objectively evaluate negative and constructive feedback, accept what works, put it into practice and leave the rest of it behind. If an injury or incident occurs or if one of our sites fails an audit, we cannot take it personally. We must be objective to determine where and how the breakdown occurred.

They put away their phones. Nothing turns someone off to you like a mid-conversation text message or even a quick glance at your phone. When genuine people commit to a conversation, they focus all of their energy on the conversation. You will find that conversations are more enjoyable and effective when you immerse yourself in them. When you robotically approach people with small talk and are tethered to your phone, this puts their brains on autopilot and prevents them from having any real affinity for you. Their genuine interest in other people makes it easy for them to ask good questions, often questions others love to answer.

They are not hypocrites. Genuine people practice what they preach. That is not usually a problem for most of us, but what happens when a senior leader walks through the shop without hearing protection? Do we give them a pass? They do not tell you to do one thing and then give a pass to someone else. We lose credibility with an employee really fast by having a double standard.

Of course, we can simply cut to the chase and demand our employees follow procedures or suffer the consequences (usually disciplinary action). That is an authoritarian management style. The funny thing about this leadership style, it works, that is, if you are satisfied with minimum standards because that is what you will get. Nor are you likely to receive any discretionary effort on the employees' part. Forget new ideas coming forth. Creative problem solving is not happening. A command and control leadership style in emergency situations is appropriate, even recommended. As an everyday management approach, not so much.

Whether you are a one-person safety department or have direct or dotted line reports, or you supervise employees or manage an operation, you are the face of safety and health (and environmental for many) at your organization. Employees "buy" you before they buy your procedures or training or anything you are selling, for that matter. Whether you realize it or not, you are branded. Everything you say and do, how you look, how you sound and the words you use are part of your brand. It all contributes to what people think of you. If you are not selling yourself, you will have a tough time selling safety. To sell safety to a front-line employee, you must reinforce your brand. You do that through the leadership characteristics just discussed and by personalizing what you do as much as possible.

The remaining chapters cover methods for personalizing safety. Effectively selling safety to the front-line employee is about influencing behaviors by making safety personal, for yourself and for them. When we make safety personal, we create an environment where they comply with procedures because we have created a connection between safety and what is most important to them. Making safety personal requires three key attributes:

1. Vision – where are you going and will they follow?
2. Knowledge – what do you need to know to get there?
3. Heart – lead with your heart.

Each chapter includes three practical actions you can implement today to sell safety, to influence front-line employees to go beyond compliance. Think of these actions as cafeteria-style plates. You can do all three or start with one you believe will work for you, but start. If you only have a few minutes, start with one you know you can complete in short order. Planning for and capitalizing on quick wins is the start of success. Maybe there is a combination you feel will work or fit with existing activities. Just start.

Three Things You Can Do Today:

1. View safety as two components. The art and the science of safety. Identify one leadership characteristic to focus on in the next month.
2. Read an article to build your technical expertise. Your credibility is based on your technical knowledge. I suggest the American Society of Safety Professional's Professional Safety Journal for starters.
3. Read an article on selling, communication, presentation or speaking to build your ability to influence others. I recommend starting with Dale Carnegie's book *How to Win Friends and Influence People*.

REFERENCE

"Fatal work-related falls to a lower level increased 26 percent from 2011 to 2016: The Economics Daily: U.S. Bureau of Labor Statistics," 2018.

SECTION 2

MAKING SAFETY PERSONAL

7

FINDING YOUR VISION

I can teach anybody how to get what they want out of life. The problem is that I can't find anybody who can tell me what they want.

– Mark Twain

Selling safety starts by making safety personal. Making safety personal starts with a vision, your vision, for safety. Having a vision is the equivalent of a growth mindset. A vision lays the groundwork for defining principles, goals and objectives. Establishing short-term goals gives employees a sense of ownership and control. When those goals are achieved, an opportunity to celebrate is created, thus reinforcing team spirit. A vision is a starting point for making employees feel like they are part of something bigger, and safety is the perfect platform.

Vision provides the destination; objectives are the roadmap. Effective strategy gives direction to everyday jobs, but it starts with a vision.

When we think of vision, words like inspiration, direction, enthusiasm, insightful, encouragement, even courage and resolve come to mind. We might think of famous people or leaders who had a particularly inspirational vision like President John F. Kennedy's "We will send a man to the moon and bring him back" or Martin Luther King's "I Have a Dream" speech or Winston Churchill's first speech as the British Prime Minister at the onset of WWII "I have nothing to offer but blood, toil, tears and sweat." They inspired people to action through their vision.

In 1904, Henry Ford had a vision to put an automobile in every home and founded the Ford Motor Company. He created the assembly line and gave his workers every weekend off. Why? So they would have leisure time to drive his car. The assembly line became a means to achieving his vision. Ford Motor Company went on to become one of the largest and most profitable companies in the world. It was also one of the few companies to survive the Great Depression. By 1929, Ford was producing 1.5 million cars, and it all started with a vision.

Henry Ford was a visionary leader. By being a better leader, a leader with a vision, we can be a better salesperson and influencer.

But we are talking about safety here. What could be more important than the personal well-being of each employee? So, why is a vision needed? Many leaders believe that if the reason for acting is important enough, like avoiding an injury, they will do the right thing. It is just common sense! Not necessarily so. Unfortunately, we often take that approach and, after some training, hope for the best. That is not how leadership works. That is not how selling works and that is not how people work. People follow worthy leaders who champion worthy causes. We can all agree that the safety and health of our employees is a worthy cause.

You are a leader regardless of your position. You do not need to be in a position of authority to be a leader. We all listen and look to our immediate boss for clues on how to act or respond to situations. Your employees will look to you as a leader. To be a leader is your choice. To be a successful leader you need a vision.

As a leader, you have an opportunity through your vision to establish a direction and enlighten and inspire employees to take action. When you find and share your vision with passion, you energize and mobilize a team to work hard together. A vision gives people the power to resolve questions before they are asked; to have a vision is to have a proactive approach to safety management. When an employee reaches a decision point to take the shortcut and get the job done or to take a minute to complete the pre-op inspection, that decision will already be made. A strong vision effectively communicated eliminates the need for the employee to make that decision because it was already made through your vision. A vision sets the standard for all employees to follow. A vision helps guide your actions, what you say and do.

WHAT IS A VISION?

According to Merriam-Webster, vision is defined as (a) "the act or power of imagination." (b) "The mode of seeing or conceiving." Can you imagine an injury-free workday? How about a work environment where employees look after each other's safety and well-being? If safety is your calling, your passion, or you at least have a strong interest, then you have a vision and just need to articulate it.

According to BusinessDictionary, "A vision is an aspirational description of what an organization would like to achieve or accomplish in the mid-term or long-term future. It is intended to serve as a clear guide for choosing current and future courses of action."

A vision statement has more to do with the future and really describes what an organization, department, team or individual person plans to be in the future. A vision is meant to inspire and motivate. It also clearly demonstrates an organization's goals to stakeholders (employees, sub-contractors, customers, investors and visitors).

Of course, it is not that simple. A vision does not just happen. It has to be developed and nurtured if it is to take hold. Many other factors must be considered to make your vision wildly effective, but most important is "buy-in" from the employee group. In John C. Maxwell's book *The 21 Irrefutable Laws of Leadership*, it is law #14, "The Law of Buy-In." Ask yourself, "Have I given my front-line employees reasons to buy into me? If the answer is yes, they will gladly buy into your well-articulated vision." If you have not built credibility with your front-line people, your vision will not matter. Safety is a worthwhile cause, but as a safety manager, operations manager or supervisor, we like to believe we have credibility based on our position. Not necessarily so. More and more consumers are making purchases online. However we still buy many big-ticket items in person. Think about the last time you bought a car, furniture or appliance. You likely did not buy it based on features or benefits. You likely bought it based on how well you liked the salesperson. Our vision works the same way. The employee group must like us and trust us first before they will decide to buy our vision (Maxwell, 1998, p. 146).

A vision statement should not discuss your present state but what you want the organization, department, team or yourself to be and how you want to be viewed. To be effective, your vision should be clear, optimistic and, of course, realistic. An unrealistic vision statement will not be motivating or inspirational but would end up being comical, especially if the organization has an injury rate well north of the industry average. For example, saying that the company will eliminate all injuries in one year would be unrealistic (unless of course you truly believe it and can make a case for it).

When considering your vision statement, it becomes important to differentiate between a mission statement and a vision statement. The key aspect to remember is the current vs. future context. A mission statement is where you are today and why you do what you do; a vision statement is where you are going and how you want to get there. While this distinction may seem trivial, it can be very important to ensure that the mission and vision are clearly defined, aligned and ultimately communicated to the people who need to know it.

Your organization may already have a mission statement. If so, you can leverage it by aligning your vision statement with the mission. A well-positioned safety vision can support your organization's mission and vision. The best vision statements do not always come from an off-site, all-day team meeting designed to "come up with" a vision statement. It can, but in my experience some of the best visions tend to evolve over time. I supported a senior executive at a Fortune 200 company. This executive hosted a monthly safety call where managers reported details of their injuries and corrective actions from the previous month. This executive started asking one question of each manager "Was that injury preventable?" Of course, the answer was always "Yes." The question eventually evolved into the slogan "Every Injury is Preventable." The slogan eventually evolved into a vision statement, "To Create an Environment Where Every Injury is Preventable." There were those who believe it is impossible to prevent every injury. This leader made them believe it was possible. He knew there was a direct correlation between possibility thinking and a person's effort level. The results were impressive although his vision could not take full credit. It did pave the path for future safety initiatives.

This brings us to another important distinction, the distinction between a slogan and a vision statement. The safety world is full of slogans; "Safety is #1," "Safety is Priority 1," "Target Zero." You see them on posters, banners and T-shirts. They hang in the breakroom, lobby or shop floor for all to see. That is where they hang and collect dust and are quickly forgotten. They make managers feel good about themselves because they believe by hanging these banners, they are effectively managing safety. These are worn-out slogans that eventually become obsolete. It is best simply to banish them to the trash heap. To be clear, it is not the slogan or statement, it's whether we bring it to life or not. Many companies have successfully brought these slogans to life. Successful leaders bring slogans to life by:

- Never missing an opportunity to talk passionately about the slogan, in person or through various means of communication.
- Defining the slogan in operational terms (more on both in Chapter 7).

A vision is something you feel in your heart, something you believe in. It is something you are passionate about. When the most successful leaders talk about their vision, and especially progress toward their vision, they do so with energy and enthusiasm.

DEVELOPING YOUR VISION

With the understanding of the distinction between mission statement vs. vision statement, and the difference between a slogan and a vision, here are some guidelines for finding your vision. Remember your goal is to inspire action through a common or shared vision.

- Establish a vision ledger and start by simply recording your general thoughts about safety and why it is important to manage. Hopefully, your general thoughts about safety are positive. If they are not, i.e., "safety is boring," then you still have a starting point. Think about what you want to accomplish, who you want to influence and how you want to go about it. Think about the people who benefit from a strong safety culture and how they benefit. You need everyone to see via your vision, the ultimate purpose of a safe operation and that creating a safe operation is a noble cause. Imagine the possibilities. What are your values? Your principles? What is most important to you? Your vision should reflect your values and principles.
- Discuss safety with front-line employees. What does safety mean to them? What is their role in creating a safe work environment? Find out what is important to them and the role safety plays in helping them get what they want most. Their opinion will almost certainly provide some insight into the direction you want to go.
- Review your organization's mission and vision statement (if one exists) and any supporting documentation like annual reports or shareholder reports. Is safety addressed? If so, what does it mean to you? Can you build off of it? Are there words, phrases or themes that grab your attention and can be applied to safety? Can you align your thoughts with it? If safety is not addressed, can you interpret or apply certain aspects to safety? Review the existing safety policy or slogans (if one exists). Is the policy common knowledge? If not, what will it take to bring it to life?
- Discuss safety with your manager, executive leader, human resources representative, training manager or other functional

leader. Ask them about their interpretation of the organization's vision and mission statement and if safety is stated or implied. Ask them about their responsibilities and accountabilities for safety. How would they describe the existing safety culture? How does it compare to your assessment of the safety culture? Ask them where they see the organization in one, three or five years. Human Resources and training departments in particular can directly impact safety and should be a key target for a discussion. They may also be your key allies in communicating your safety vision.

- Is there a gap between leadership's perspective and the front-line employees' perspective? How big is the gap? Do they agree on any aspect? Can you pinpoint specific aspects where each group differs? Is there a recurring theme?
- Is it achievable? Realistic? An unrealistic vision statement will not be motivating or inspirational, i.e., saying that in one year, the company will eliminate all injuries may be a worthy goal, but many may see it as unachievable.

By now, a few thoughts should be rising to the top. If so, and you have at least a draft vision statement, begin to map out your intended path. What actions are required for you to reach your destination? List three to five actions you can take to support your safety initiative. You may not have landed on a clear vision yet and that is ok. It will come.

Were you able to identify some guiding principles, themes or at least a general direction? Consider these potential scenarios:

Scenario #1: You find out from reviewing your organization's vision and mission statements that safety is not a shared value and it is reflected in your discussions with front-line employees. Your vision may begin to revolve around "sharing" safety and the importance of sharing values such as the safety and health of your coworkers, friends and family.

Scenario #2: You find that safety is treated or managed separately from other aspects of the business such as finance, human resources and production. Safety is an add on and only addressed when time permits. If that is the case, your vision may evolve around "integrating" safety into all aspects of the job; "Safety is integral to our success." Your discussion will involve integrating safety into existing procedures or training, or new employee orientation.

One organization I worked with mentioned the importance of "people" on several occasions in their vision and mission statements. While safety was not specifically mentioned, we applied safety as a supporting aspect of the corporate vision statement. The safety vision evolved to become "No one gets hurt." All communications about the safety vision referenced the emphasis on "people" in the corporate vision and mission statements.

Scenario #3: It is not unusual to find that regulatory compliance is emphasized in the vision and mission statements. Compliance, while very important, is often characterized by command and control, inspections, audits and disciplinary action, all of which tend to evoke thoughts of minimum standards at best and negativity at worst. If that is the case, your vision may evolve around commitment and cooperation rather than compliance. Your vision should inspire, not intimidate. It is difficult for anyone to get inspired by a vision to be in compliance with all federal regulations.

There is no real timeline for developing and establishing a vision. As I mentioned earlier, some of the best vision statements evolve over time and often are based on a collective experience. Most important is to put it in writing and start talking about it. Talk about it with your team, your supervisor or manager, your friends and family, your sub-contractors and your vendors. Just start talking about it and never miss an opportunity to share your vision or thoughts about your vision. You will find that your thoughts will evolve, and your vision will begin to clarify.

Do not wait for the perfect vision to start sharing it or talking about it. If you wait until you feel it is perfect, it may never see the light of day. You can and probably will completely change your vision in particular if you feel you have achieved your vision or maybe the culture has improved or changed. You may also have a specific vision as it relates to a project or task.

OBSTACLES

You will face some powerful operational headwinds. Do not get discouraged! Here are a few of the most common obstacles.

1. Corporate culture vs. sub-cultures and the difficulties we face in establishing a safety culture that is not supported by the corporate culture. Organizational culture influences how an

employee acts and can be an impediment to an employee working in a safe manner. As a front-line employee, I experienced
a culture that in effect valued timely operational performance
over employee safety. The results were predictable. A safety
vision will always be dominated by the corporate culture, but
it gives you a fighting chance to make at least a small difference in an area where you have influence. You might not be
able to change the organization's culture, but you can influence the character of the individuals on the front line through a
well-articulated safety vision.

According to Edgar Schein (2004, p. 198) in his book
Organizational Culture and Leadership, "The success of the
organization depends on the knowledge, skill and commitment
of the front line employees." If you provide the shared vision,
you will get the commitment.

2. Employee perceptions about the organization and safety in general can affect your level of success. Perception can be based
   on the employees' opinion of management's view on safety, the
   previous supervisor's view on safety or their experience at a
   previous job. Be genuine in your discussions about safety and
   they will see you in a different light.

3. Operations managers, supervisors and safety professionals are
   generally oriented toward the needs at the moment, how to get
   the task or project done and how many people are needed to
   cover the shift weekly or monthly inspections. We put out daily
   fires. We should and do spend a majority of our time on the
   immediate and near future; however, a few minutes a day or
   week considering the long-range implications and where you
   want to be in one year, three years or five years helps to shape
   your plan for today and tomorrow. If you want to be a leader
   that is followed, be forward-looking.

Statements from executive leaders, banners and rewards alone do not
inspire people. Uninspired people do uninspired work. They meet
minimum standards. They do not give of their discretionary effort.
On the other hand, passionate people are involved, they give extra
effort. In the context of safety, they look out for each other. Where
does that passion come from? From your vision, goals and ideas that
provide inspiration to go above and beyond.

I know what you are thinking, because it is what I thought for
many years. "I barely have time to keep up with the daily routine stuff

and the fires that pop up constantly." I will say this: if you want to inspire your employees, if you want to make safety personal, finding and communicating your vision is the place to start. Start small, but start today.

The idea for this book started with a presentation I gave at numerous conferences titled "Selling Safety to the Front-Line Employee". My vision was simply to teach how to sell safety. It evolved into something akin to sharing my experience. After considerable thoughts, discussions and questioning my motives and what I felt passionate about, I landed here: "I will be a compelling force behind which everyone will be inspired to make safety personal." In short, "Make Safety Personal."

Three Things You Can Do Today:

1. Start a vision log and make your first entry. Document your initial thoughts. Schedule 10–15 minutes, at the same time each week to work on your vision. It can be used to talk to someone or document thoughts or observations.
2. Review an existing company document such as the company annual report, mission, vision or safety policy if one exists.
3. Share your thoughts with one person or create a vision board and ask others to contribute. Ask a front-line employee to document the importance of safety.

REFERENCES

Maxwell, John C. (1998). *The 21 Irrefutable Laws of Leadership*, Thomas Nelson.

Schein, Edgar H. (2004). *Organizational Culture and Leadership*, John Wiley & Sons.

8

COMMUNICATING YOUR VISION

The primary goal for any safety management system to be successful is to clearly articulate the overall vision and mission, safety policy statement and stated goals and objectives to everyone in the organization.

– James Roughton and Nathan Crutchfield (2014, p. 67)

You cannot effectively sell safety without being a good communicator and that means articulating a common vision that inspires employees to take action. Name a great leader or a leader you admire, and chances are pretty good they are excellent communicators. They articulate their message in a manner you understand and get excited about. We are attracted to people who can communicate with enthusiasm, what they believe, where they are going and why we should join them. They make us feel like part of something bigger, they make us feel like we can make a meaningful contribution that will have a bigger impact than if we were working alone. They can inspire us through their words and actions. Our job is to inspire front-line employees to go beyond compliance, to recognize and report hazards, to value not just their personal safety, but their coworkers' safety.

The key is they must know it is important to their supervisor. You accomplish this goal by creating a shared vision. The more you talk about your vision the closer you come to achieving a shared vision. So, keep talking! The most influential leaders never miss an opportunity to share their vision in a passionate and enthusiastic manner. When they bump into someone unexpectedly, they have a 30 second speech on the ready to take advantage of those few seconds. They look for and even seek out opportunities to talk about their vision and what it means to their audience, organization or team.

There are three (3) components in the communication process:

1. Sender
2. Message
3. Receiver

Effective communication starts with the first component, you, the sender. Anyone can send a message, but unless the sender has integrity and is believable, the message will fall on deaf ears. If you do not believe the messenger, you will not believe the message. You cannot go to the next step until you are rock-solid in the integrity, credibility and trust categories. If you are not sure, ask yourself these questions:

- Do employees seek your advice or confide in you?
- Are employees eager to participate in your safety committee or volunteer for projects you suggest?
- Do they report safety concerns to you or contribute to safety discussions in a constructive manner?
- Are they willing to give extra effort when the operation or situation calls for it?

If the answer is yes to any of these questions, you likely have some level of trust, meaning your messages are also likely to be received, understood and believed.

The second component is the message. Too often, we get lazy when discussing safety and it results in a limited vocabulary and lack of creativity. Often, the receiver of our message gets bored and loses interest. In other words, it is too easy for us to simply say "be safe," "be careful," "pay attention" and "use common sense." When we do, our message gets lost.

Try this exercise with your team. In a group setting, ask each member to define "safety." If you have 10 people in your group, you will likely get 10 different responses. The takeaway from this exercise is that by using the word "safety," you are saying something different to everyone. Vision statements can be just as abstract. To send a clear message, safety and your vision must be defined in the frontline employees' terms, meaning in operational terms that all receivers understand and can act on. According to Joel Kurtzman (2010, pp. 101–102) in his book *Common Purpose*, "The right environment promotes an atmosphere where the direction is clear and where you can translate that vision into doable things – things you can give to your team that they can actually go out, do and accomplish."

We should do the same for safety. Rather than saying "be safe," define "be safe" in operational terms. For example, if you anticipate more than usual forklift activity, rather than simply saying, "Be safe in the warehouse today," you might say, "We have a shipment coming in today, so forklift activity will increase. Stay in the pedestrian zones while in the warehouse."

Inspiring leaders "use metaphors and analogies; they give us examples, tell stories, and relate anecdotes." Metaphors are every-where. Metaphors trigger the right hemisphere of the brain, a critical component for persuasion to occur. They include combat and sporting metaphors as well as spiritual metaphors (Kouzes and Posner, 2017, p. 131). "Safety is a team sport" or "Everyone is on the safety team" are sports metaphors.

Create word pictures by using images. Another good exercise is to ask your audience "What is the first thing that comes to mind when you hear the word safety?" They likely will struggle for a minute before calling out images of safety. They might mention specific personal protective equipment (PPE) such as a hardhat, a reflective vest or maybe machine guards or lockout tags. They probably will not think of injury or accident rates. Use words and images in your communication that evoke sights, sounds, smells and tactile feelings. "People aren't persuaded by the facts as much as they are by the emo-tions, feelings and images behind those facts," according to Posner and Kouzes (2017, p. 127) in their book *The Leadership Challenge*.

Translate your vision into operational terms. Let's return to our example vision "Safety is a Team Sport." How could this vision be defined in operational terms? What would it look like from your own perspective? In the previous forklift scenario, your message could emphasize the teamwork between pedestrians and forklift opera-tors. What does it look like? It looks like eye contact between the operator and pedestrian or hand signals indicating you are clear. That takes teamwork! The kind of teamwork that prevents struck-by inju-ries involving forklifts. Now we are defining safety and your vision in operational terms that your receiver can comprehend. Further, by defining safety based on your vision, you begin to create a shared vision (Figure 8.1).

To take the sports metaphor another step, everyone wants to know the score if they are watching a sporting event. The same applies to

Figure 8.1 Safety is a Team Sport. Illustration by Masha Balac.

safety. Everyone should know whether the team or organization is improving, but using numbers without context can be confusing. Limit the use of numbers to one key number. If you do use a number, especially if you use a rate such as the Total Recordable Injury Rate (TRIR), provide some context behind the number. For example, if you have 100 employees working one full year and a TRIR of 5.0, that would be the equivalent of 1 in 20 employees being injured seriously enough to require medical attention beyond first aid (more on the importance of context in Chapter 10). Finally, stay away from clichés such as "it could have been worse," "accidents happen" and "we need to stay focused" in your messages. They are overused and lack meaning.

The third component of communication is the receiver. Who needs to hear your message? The receiver is the one you want to respond to your message. This list includes all front-line employees or at least those in your sphere of influence. Your message may vary somewhat depending on your receiver. Using our example vision, "Safety is a Team Sport," the new employee may need to hear why teamwork is important to ensure a safe work environment. For any junior- or senior-level front-line employee, emphasize the personal connection to safety. An operations supervisor may need to hear how teamwork in safety can improve efficiency and help manage workloads. Regardless, the theme of "Safety is a Team Sport" should be consistent.

MEANS OF COMMUNICATION

The most common platforms for communicating a safety message tend to be safety meetings, shift briefings or toolbox talks. Employees expect to hear the "safety" message during these events, but we often overlook other platforms that can be used to communicate a safety message. Some platforms can be highly effective. I worked with a Vice President early in my safety career who was fond of saying, "Tell them 10 times, 10 different ways and they will remember." I have come to call it the 10 × 10 communication rule. Different methods help not only to get your message out and to reach more employees, but keep your message fresh. Some methods have higher retention than others.

The following are opportunities we should not miss and examples of how to use each means:

1. Pre-task analysis – What does this task require to ensure your safety vision is achieved? It could be a specific line item or a general discussion among the team or crew during the pre-task analysis.

2. End of shift reports, after-action reports or post-incident review – After action reports identify what worked and did not work and lead to opportunities for improvement. Further, it is an opportunity to identify specific actions taken or could have been taken to achieve your vision of safety or safety goals. It could be expressed in a question, "What did we do (or could have done) today to encourage teamwork and create a safe work environment?"

3. Reward and recognition – What are the specific behaviors that embody your vision? Communicate and reinforce those behaviors through recognition. Make a point of focusing on one behavior each month. For example, using the sports analogy, your communication might read, "Safety is a Team Sport. That means we use team lifts to move stock that weighs over 24 kg (53 lbs). Over the next 30 days, we will conduct observations and recognize anyone using the team lift."

4. New employee orientation and on-boarding – Set expectations up front by defining your vision in operational terms for every new employee. The best example I have seen is an employee "safety pledge" included in the new employee orientation. The "safety pledge" defined the company's vision for safety

in 14 clear and actionable terms and was signed by the new employee. The safety "pledge" was included in the safety orientation handbook, which further defined and illustrated specific behaviors. The "Safety Pledge" introduces an interesting and potentially impactful opportunity. By signing and posting the "pledge" in a public area, it becomes a visible sign of commitment. A public display of commitment is a powerful influence on the new hire and other employees. Another option if there are several employees going through orientation and they all plan to work in the same area is to create a poster of the commitment. The "pledge" should be reinforced on a routine basis. Otherwise, it runs the risk of becoming obsolete. The same can be applied to sub-contractors and visitors.

5. Employee evaluations – What actions did the employee take to achieve a safe operation or further the safety vision during the evaluation period? At the very least, evaluations are an opportunity to reinforce your vision for safety verbally to the employee and in writing. It is easy to focus on areas of improvement; however, looking for and documenting positive behaviors help ensure those behaviors are repeated. Hopefully, your employees hear about your safety vision more frequently than during what is typically annual evaluations.

6. Refresher and recurrent training – The instructor or trainer should discuss how the training helps achieve your safety vision. Avoid explaining that training is required by the Occupational Safety and Health Administration (OSHA) or other regulatory agencies. Otherwise you will get a minimum standard at best and at worse reinforce negative connotations about safety.

7. Job descriptions – These should include responsibilities associated with the safety vision. Set expectations before the employee is hired. The most successful safety cultures begin here.

8. Handwritten notes – Few people seem to write by hand today, but it can add a very personal touch. One of the best examples of messaging using handwritten notes I have seen was from an executive chef of a high output kitchen who had a clear vision for kitchen safety (Slip, Trip and Fall Free Kitchen, Cut Free Kitchen) and penned messages on the recipes at each workstation.

9. Employee bulletin boards – Sign and post a copy of your vision on the company bulletin board. Post handwritten copies at various places at the worksite or shop floor. Can you use a picture or

image to depict your vision? Attention and retention go up with visual aids. Remember "show and tell?" I often see bulletin boards covered with outdated memos. Some several years old. Be sure to move or update these copies periodically, otherwise they will begin to blend into the woodwork. Don't wait until you have honed your vision. This can be an opportunity to ask employees to add their comments in writing.

10. Elevator speech – A term used to describe a brief speech that outlines an idea. The name comes from the idea that the speech should be delivered in the short time period of an elevator ride, usually 20–60 seconds. If you are given one minute to address a training class or new employees during orientation, ask yourself "How best can I use that time?" Having an elevator speech that captures the essence of your safety vision uses that time wisely and can provide another avenue for communicating your vision. Vary your elevator speech based on your audience.

For a shift briefing or introducing yourself, an elevator speech might sound something like this:

"I am your safety manager (or supervisor), but I used to work on the floor like you. Every day that we committed safety infractions, we failed to follow training techniques and put each other at risk. Then one day, a coworker suffered a serious injury. The injury impacted not only his life, but his family's life. It also impacted his coworkers because they felt they contributed to the incident by their at-risk behaviors and failure to speak up. Because of that incident, I joined the safety department where my job is to ensure you have the right tools, knowledge and skills for the job, and we are identifying and correcting hazards before someone gets hurt. I encourage you to do your part and when you see a condition that causes you concern, report it to your supervisor or me immediately."

For a new hire orientation class, an elevator speech might sound something like this:

"I am your safety manager (or supervisor). Can I ask you a question? What would your life look like if you suffered a serious back or hand injury? Your job involves working with machinery that can cause serious injuries when procedures and training techniques are not followed. We share a common goal, to ensure no one gets hurt today. To achieve this goal, my job is to ensure that the right controls are in place and you are armed with the knowledge, skills and tools to prevent an injury. To achieve our goal, we need you to do three things:

1. Follow procedures and training techniques.
2. When in doubt, stop and ask your supervisor or contact me.
3. Report and correct hazards.

If there is anything I can contribute to help you achieve our shared goal, please let me know."

Another example:

"I am your safety manager (or supervisor). I believe every employee deserves to work in an environment free of recognized hazards. To work in an environment where you can do what you do best without the worry of being injured. My vision statement is simple, "Everyone Goes Home Safe Today." To achieve my goal, I do three things:

1. I ensure hazards are identified by conducting observations and inspections,
2. I ensure that procedures and training address hazards, and
3. I ensure safety concerns and corrective actions are closed in a timely manner.

The safety committee plays an important role. I ask that you read the minutes from each meeting, report hazards to a safety committee member and consider joining the committee. We all have an important role to play that contributes to a hazard-free work environment. When we work together, everyone goes home safe today."

In keeping with the sports analogy, an elevator speech might sound something like this:

"Achieving excellence in safety takes a team effort much like a football team that wins the championship. We win when everyone goes home today to enjoy their hobbies or playing with their kids. Working as a team means we look out for each other, we identify and correct hazards, report hazards and near-miss incidents, and follow all training techniques. Above all, when in doubt, stop and ask a teammate. My role as your safety manager (supervisor) is to provide you with the knowledge, skills, tools and equipment to identify and correct hazards before an injury occurs. We all have a role to play on the safety team and when we play that role, we win."

Notice the infrequent use of the word "safe" or "safety" in each elevator speech and mostly in the title "safety manager." Refrain from over-using the word "safety;" otherwise your message begins to become cloudy.

The means of communication are only limited to our imagination. Text messages are being used more frequently today, but should not replace more personal communications such as handwritten notes. Further, you should not limit the avenues you use, but may select the ones that have the most promise. For example, including your vision messaging in new employee orientation seems an obvious choice, but without including your messaging in recurrent or refresher training, your message may soon be lost. Remember, the goal is to create a "shared vision." Reach as many employees as possible and at different stages of their employment. Use pictures, images, props and videos to improve retention. Most of all, keep talking!

There is one method that stands out above the rest and is perhaps the most important for communicating your safety message: utilizing allies. There is no better person to do that than an informal leader among the front-line employee group. These are your allies. The informal leaders are the employees who others look up to, listen to or go to for direction or instruction. The most influential informal leaders are the ones who have an interest or at least understand the importance of a safe and healthy work site. Find out who they are and share your vision. Recruit them to be a spokesperson for your vision. Why? Your message coming from a coworker, especially one who is an informal leader, is more powerful than your message coming from you. Most front-line employees expect their leader to provide direction and to talk about what is important. An informal leader sharing the same message automatically gives your message more credibility. Making an ally out of an informal leader is an effective means for creating a shared vision. In addition, your allies can reach more people than you can alone.

Using allies is especially effective in large, geographically dispersed organizations where simply getting one message out to all front-line employees is a challenge. You might start by asking your informal leader to open the next shift briefing or meeting with a "safety moment" or "safety share." Another option is to ask them to work with you to develop your vision or ask for their opinion. Most people are flattered when someone thinks enough of them to ask for their opinion (more on allies in Chapter 8).

IT IS ALL IN THE PRESENTATION

There are two types of speakers: those who get nervous and those who are liars.

– Mark Twain

The time comes and you are asked to formally present your vision for safety (if it hasn't already). Do you get so excited that you immediately start taking notes about your message, or are you like the rest of us and get a little queasy at the thought of public speaking? For many people, it is the one thing they fear the most. Most people just do not like to give formal presentations or maybe it is the fear of making a "you know what" of yourself in front of other people. If you are one of those people, there is good news. Public speaking and presentation skills are learned skills. If you want to be wildly successful at selling safety and influencing front-line employees, this is another area you need high scores. You need to be able to deliver an engaging presentation that sticks, whether it is a one-on-one conversation, an informal toolbox talk with five employees, or a formal presentation to the operations team of 125. Each situation is an opportunity to sell safety, to sell your vision and to influence others to follow.

Being comfortable speaking in front of a group is NOT the same as being a good presenter. I have heard many people claim to be good speakers because they were comfortable speaking in front of groups. They are, but their presentation skills send a completely different message. It takes knowing and tailoring your message to your audience. Having a vision, even a rough idea, and a message to tell is a great starting point and gives you confidence. In addition, you need to know how to use your voice and body language to your advantage. If you are using PowerPoint or other slide presentation program, you need to know presentation design. I emphasize "if" here because some of the most powerful presenters do not use slides.

Start by knowing your audience. Knowing your audience helps you craft your message. A strong message for the right audience is essential to a successful presentation. Ask yourself first: What do you want them to know? How do you want them to feel? What do you want them to do? Are there supervisors in the audience? Managers? Contractors? Vendors? What is my core point? Or put it this way, as Gar Reynolds says in his book *PresentationZen* (2012, p. 63), "If the audience will remember only one thing (and you'll be lucky if they do), what do you want them to remember?" And why does it matter? or simply stated, so what?

We actually communicate in three different methods: verbal, vocal and visual. In other words, by what we say (verbal), how we sound (vocal) and how we look (visual) when we speak. Studies show that when our words are not consistent with how we sound or how we look, the audience is more likely to believe the messages our body language is sending.

Perhaps you have attended a safety meeting that went something like this: the facilitator walks to the front of the room with slouched shoulders, hands in pocket, weight shifted to one side and in a low voice opens the meeting by thanking you for attending, explaining how important safety is and promising not to run the meeting too long because well, safety is kind of boring.

According to a study by Albert Mehrabian (1971, p. 43), in this situation, when what we say is not congruent with how we sound or how we look, 55% of the time the audience will believe how the speaker looks, 38% of the time they will believe how the speaker sounds and only 7% of the time will they believe what the speaker actually says. With these facts in mind on the importance of nonverbal communication, here are guidelines for a successful presentation.

Verbal (What You Say)

Start by being clear about your message as it relates to your audience. Note message in the singular, not plural. Attempting to communicate multiple messages confuses the receiver. If the receiver is not sure what you are saying or confused about the key message, they will default to the nonverbal cues. Decide on the message and action you want them to take. Refrain from non-words and fillers like "umm" or "uh." Mine is "so." I tend to use the word "so" as a filler or to bridge one thought with another. I am still working on completely eliminating it from my presentations.

Vocal (How You Sound)

When you speak, vary your pitch and delivery speed. Raise your pitch when you want to make a key point. Use pauses for emphasis and drama. A 3–5 second pause allows your audience to absorb your message. Silence is good, but too often we get uncomfortable with silence or dead space and feel the need to fill it in. Pauses are your friend when speaking, so use them. When your audience hears you speak, they should detect enthusiasm and passion in your voice. No one likes listening to a monotoned, unenthusiastic speaker, but add voice inflection, varied delivery speed, pauses and passion and they will perk up.

Visual (How You Look)

Our bodies speak a language all their own, and they do not lie. Your body language has likely become so ingrained and habitual, to the point where you might not even think about it. If that is the case, it is time to start, because you could be sabotaging your message.

Start with your stance. Maintain good posture, relax your knees, balance your weight between both feet and keep your feet hip-width apart. Chin up. Shoulders back, do not slouch. It is called the "power stance" because it projects an image of confidence.

Refrain from pacing. It is a nervous habit for many speakers (something else I am working on). I often hear a speaker make an excuse by saying "I like to move around." Moving is good if it is done with purpose. Pacing back and forth in front of the room or from front to back is not moving with purpose. It only ensures you will aggravate your audience or worse put you in a position of speaking to the back of their heads. Influential speakers move with purpose. For example, take a step or two toward your audience to make a key point, then step back and pause three to five seconds to allow your point to sink in.

Make an emotional connection with your audience by holding eye contact for two to three seconds. A good method is to deliver a full thought or sentence to one person, then shift your focus to another person. Your audience wants to feel like you are talking to them as an individual. Skimming across the top of their heads, a habit of unprepared speakers, ensures that you do not make a connection.

What you do with your hands sends a message as in the previous scenario. Use open hands and palms to create trust. Do not point or use exaggerated gestures. Be aware of your facial expressions. Inconsistency between your words and your facial expression causes people to sense that something is not right.

Use props or other visuals to help your receiver visualize and remember your message. Visuals dramatically improve message retention. In one presentation to a group of chefs and cooks, I used a kitchen timer to illustrate just how quickly a safety observation can be conducted. I set the timer for 45 seconds and conducted a mock observation. I finished with several seconds to spare. Using a prop and providing a visual of an observation improved understanding and retention of my message.

It is almost impossible to pick up on bad habits without seeing yourself. The best process for correcting bad habits and becoming an effective speaker is to video yourself. It is difficult to watch yourself on video or listen to audio, but you will quickly identify areas for improvement. You will notice those non-words and fillers you had no idea were coming out of your mouth.

Are you still nervous before you address an audience or your team? Many experienced speakers still get nervous. Here are some methods I have used to help reduce anxiety and nerves.

- Work the room before your appointed start time. Consider yourself on stage as soon as the first person walks in. Greeting employees as they arrive informally begins your presentation, which makes the transition to your formal presentation smooth.
- A few minutes before the presentation, conduct some breathing exercises. Take a few slow, deep breaths, hold it for a second and slowly exhale.
- Stand in a "superman" pose immediately before the presentation. Stand straight with shoulders back, feet a little wider than hip-width apart and hands on your hips. This pose gives you confidence and makes you feel strong. Audiences pick up quickly on a lack of confidence.
- Forget the coffee or at least switch to decaf. I usually only drink one to two cups a day, but I found out firsthand that is all it takes for me to turn a 50-minute presentation into a 25-minute presentation. I left their heads spinning and wondering what just happened.
- An important point to remember is that your audience is pulling for you. They want you to do well. They want to hear a dynamic speaker with a clear and concise message. If they come expecting another boring safety talk, then surprise them. It won't be difficult. Remember, it is up to you to deliver.

In summary, it is not always what you say, but how you say it and how you look when you say it. If your words do not agree with how you sound and look, people will believe your tone and nonverbal behavior.

Now that you have identified your audience and crafted your message, it seems to be an automatic reflex to build a presentation deck, usually in PowerPoint (PPT). Somewhere along the way, we became enamored with PPT. It was cool! PPT made all of us better speakers. Right? Unfortunately, our presentations became about the PPT deck. Has your boss ever said to you, "Build a deck for the shift briefing and send it to me so I can review it." What did we do? We loaded it with lots of nonessential information to show that we knew our stuff. At best that showed a lack of creativity, at worst it became an incredibly boring and undecipherable presentation that did not support our message.

Our slide deck should complement our presentation, not be the presentation. If your audience is reading your slides while you are talking, you can be assured of two things. One, they are not listening to you. Two, you have too much information on your slides. Create a written document to leave behind after your presentation with details if necessary. It will satisfy the intellectuals in your audience. Just be sure to hand it out after your presentation, otherwise they will read it while you are speaking. Most front-line employees probably will not be interested in a document, but have one available anyway. According to Gar Reynolds in his book *PresentationZen* (2012, p. 20), "The purpose of your presentation is to make an emotional sale, to influence the audience to take the action you are prescribing."

Besides being optional, here are some general guidelines to keep in mind if you must build a slide deck.

Create slides that reinforce your message, not repeat them. Clarity is too often missing from slide decks, so use no more than six words per slide. There are other rules for word use. For example, the 4 × 4 rule or 6 × 6 rule. No more than four bullet points and four words per bullet point per slide. They work, but I prefer the six words per slide rule. Regardless of the rule you use, overly wordy slides only ensure your audience is reading and not listening to you.

Empty space on a slide is not a bad thing. Yet we often see how much info we can cram on a slide. Do not take up important space on the slide by including your company logo on every slide. I am sure you and your audience know the name of your employer.

To boost your message, simplify your slide. Your audience cannot spread the word if they do not get what THE word is. Start by reducing your slides to the essential information.

Do not use cheesy images, clip art or poor quality pictures. They look cheap and will reflect poorly on your message. Rather than showing a bullet list of lagging indicators, show pictures or images of the primary injury drivers. Pictures, charts and graphs are more powerful than words. You can get high-quality pictures from these sites: iStockphoto.com and pixaby.com. Google Images is a clearinghouse of images, but beware of the quality and copyright ownership. Finally, when it comes to pictures and size, do not stretch pictures to fit the slide. It degrades the picture and your presentation and gives the impression that you did not want to take the time to find a picture that fits the screen.

Above all, do not read from your slides. You should know your material well enough to speak to your slides, not from your slides.

Using a quote adds credibility to your presentation. Have a member of the leadership team provide a quote and include it on a slide with their picture. Ideally, the quote will support or reinforce your vision or message. Reading a quote is the only situation where reading directly from a slide is acceptable.

Remember, the purpose of your presentation should be to make an emotional sale that prompts your audience to take action. Action based on a shared vision. Be simple, be accurate.

In summary, PPT is an optional presentation tool. Giving a presentation without a slide deck ensures the audience's attention will be on you. If you must use PPT, use it as a tool to support your presentation, not to be your presentation. Identify your core message and create simple slides that support your message. Know your material inside and out. It will reflect well on you as a credible person and put you in a stronger position to sell safety and your vision.

There are numerous books written on this topic (I mentioned one of my favorites earlier) and workshops designed to improve presentation skills and slide design. I encourage you to research, read or take a workshop to improve your skills. It is time well spent. If you have sat through a bad presentation, you know what I mean. A great vision and message can only be great if it can be communicated in an effective manner.

TELL THE STORY

Your employees may be well informed or experienced on the job. They may even be more experienced and employed longer than you. If that is the case, you cannot beg for or buy their attention. You cannot expect them to stand at attention just because you showed up and have a vision. One very successful communication tool is the use of stories.

Tell a story to support your vision or message. We care about and love stories. We remember stories. We imagine ourselves in the story and how we would react. Fortunately, safety is a wonderful platform for stories, especially success stories. Stories are the key to getting your message across. We love numbers, but we overuse them or at least do not provide context behind them. The problem is, people are not moved to action by numbers alone. Have you ever posted a graph of your injury rate that showed an increase and expected behaviors to miraculously change? When they did not, you could not figure out why. I know I did. I spent hours making colorful charts and graphs that did not have an impact on one front-line employee. If you want to sell safety, you need to evoke their emotions and you cannot do that with numbers and graphs alone.

Safety is emotional, so do not be afraid to evoke the emotion in safety. We battle every day to build a safety culture, to overcome disinterest and the "way we've always done it." Yet, everyone expects to go home safe. The reality is that some employees are injured either on the job or on the way to or from work. There is emotion in these situations. There is also the second victim phenomenon that is full of emotion. The second victims are the people who suffer as a result of someone else's serious injury or fatality. The coworkers, friends and family members. To sell safety and make it stick, we need to share these stories.

Our stories should emphasize the contrasts in safety, the before and after, not just the negative aspect. Tell the story about how the hazard was engineered out. How it was reported and fixed and the results. Tell the story of how you arrived at your vision or why safety is important to you. Even better, have someone else, maybe an employee, tell their story.

According to authors Chip and Dan Heath from their book *Made To Stick* (2008), messages that we remember have common characteristics. I adapted those characteristics here:

- Simple – What is the core of your idea or vision? "Core messages help people avoid bad choices by reminding them of what is important." Keep it simple. If every metric, policy or procedure is important, then none are important. Answer the question, "What is the one key point I am trying to make with my message?"
- Unexpected – Wake up your audience by filling a gap in their knowledge with surprising information. For example, according to the Bureau of Labor Statistics (2016), fatal work injuries from slips, trips and falls have increased 25% since 2011. Over 800 workers die every year from a fall. Tell the story of a fall incident and actions required to reduce the risk of a fall.

- Concrete – The word "safety" itself is an abstract word. Define safety, your vision and your message in operational terms that can be actioned. Refrain from using clichés and when possible limit your use of the word "safety."
- Credible – Lend credibility to your message by using an honest and trustworthy source. Quote a respected person within the company or have an employee who suffered an injury tell his or her story. "Telling stories using real people is the most compelling way," according to the authors. We often use statistics to provide credibility to our message; however, statistics without context lack value and impact.
- Emotional – Emotions, not numbers, inspire people to act. Tap into what people care most about and connect it to your message or vision. Safety is full of emotion. Think about the emotional impact of serious injuries on the injured employee, their families and other second victims. To say that building a strong safety culture will avoid costly regulatory fines may be true, but does not evoke emotions. To say that a strong safety culture helps to ensure "Joe" can teach his grandchildren how to fish does. Associate safety with what people already care about, their coworkers, family, friends, hobbies, pets, etc.
- Story – Stories provide inspiration. We are wired to tell and listen to stories. Tell success stories or stories of an incident and how it happened. According to the Heath brothers, stories act as a kind of mental flight simulator, preparing us to respond more quickly and effectively. Mental simulation is not as good as actually doing something, but it is the next best thing. The right story is like a flight simulator for the brain.

Anyone can tell a story. The challenge is to find them. If you can condition yourself to look for, listen for, identify and tell stories, you will stand out among presenters from other functional areas of the company and your message will stick (Heath and Heath, 2008, p. 213).

The best leaders never miss an opportunity to share their vision. They go even further by looking for opportunities and methods to share their message and their story. So keep your message simple and be relentless and consistent when it comes to sharing your vision. Weave your messaging into day-to-day conversations, into daily shift briefings or audit findings. Tell a story that conveys your vision. Add a personal touch with handwritten notes. Just be sure your messaging is aligned with your vision and where possible your company's vision and mission. Above all, let your passion and enthusiasm come through. You just might find that it is contagious.

You will know when your message is connecting when other employees start to use your lingo or better yet, repeat your vision.

Three Things You Can Do Today:

1. Practice, practice, practice to deliver a killer presentation. Video yourself to understand where you need to improve.
2. Use at least one new communication avenue to share your vision or message.
3. Start looking and listening for stories to tell. Stories that convey or support your vision.

REFERENCES

Bureau of Labor Statistics, Census of Fatal Occupational Injuries, 2016.
Heath, Chip and Dan Heath (2008). *Made to Stick*, Random House.
Kouzes, James M. and Barry Z. Posner (2017). *The Leadership Challenge*, John Wiley & Sons.
Kurtzman, Joel (2010). *Common Purpose*, Jossey-Bass.
Mehrabian, Albert (1971). *Silent Messages*, Wadsworth.
Reynolds, Gar (2012). *PresentationZen*, New Riders.
Roughton, James and Nathan Crutchfield (2014). *Safety Culture, An Innovative Leadership Approach*, Elsevier.

9

KNOWLEDGE, KNOW WHAT THEY WANT MOST

Karol, make sure your team completes the mandatory safety training this week!

– My boss to me

We have to complete the annual Lockout/Tagout and Hazcom training by next Friday. It is mandatory and boring as hell, I know, but we have to get it done otherwise we'll be late.

– Me, as a front-line supervisor to my team

Regardless of where you are in terms of your vision and how well you have communicated it, this chapter is about your target audience: the front-line employee. You do not need a draft vision or an idea to benefit from this chapter. In fact, this chapter may help you craft your vision.

Think of front-line employees as buyers – buyers of safety and health; and we as sellers – sellers of safety as a product or service. Dale Carnegie, in his book *Winning Friends and Influencing People*, stated it simply, "Selling is about influencing the other person to do what you want them to do. The only way on earth to influence others is to talk about what they want and show them how to get it." In our case, show them how safety can help them get what they want. To do that, we need to know what they want most.

To gain this knowledge, we must be a good conversationalist, and to be a good conversationalist, we must be an attentive listener. To be interesting, we must be interested. Interested in what they are saying. Ask questions that others will enjoy answering. Encourage them to talk about themselves and their accomplishments, their goals, their hobbies and what they like to do in their spare time. Why is this knowledge important? Because answers to these questions help us connect safety to what they consider to be most important. Once you have made that connection, you are in a position to link safety to their personal values and show them how safety can help them get what they want most.

For example, you might find out that an employee has a woodworking hobby. Now you have the opportunity to discuss the importance of pre-operation inspection of equipment, machine guarding and personal protective equipment (PPE). Make the connection between safety at work and home and how conducting pre-op inspections, using proper machine guarding and PPE helps them enjoy their hobby. Knowing what an employee wants most provides a platform for discussing hazards, connects safety to something that is important to them and shows you care.

I have often used my personal example in presentations. I enjoy umpiring Little League baseball games. As an umpire, I wear a face mask, chest protector, shin guards and protective shoes. I take care of my equipment and keep it in an easy to access location during the baseball season. From here, it is not a stretch to connect what I enjoy doing as a hobby to safety on the job, in particular to PPE. You might even ask me to address the importance of PPE at the next safety meeting and what it means to me personally.

Remember that people you are talking to are far more interested in themselves and their problems than they are in you and your problems. Be a good listener and encourage others to talk about themselves. Not only is it their favorite topic, but if you are asking in a genuine manner, it shows you care. In addition to connecting what they want most to safety, these conversations can help you begin to form or clarify your vision.

Some might think it is prying into their personal business. To be clear, I am not suggesting you engage in deceitful behavior. This will only serve to erode your trust and, more important, devalue safety and health. I am suggesting that you have purposeful and genuine conversations. Anything outside of that will be seen as an attempt to manipulate. When the purpose of your conversation is clearly to create a safer and healthier workplace, you will gain trust and respect, and the knowledge you need to connect safety to what they want most (Carnegie, 1936).

It is important to note that employees make decisions based on what they believe is important to their immediate supervisor. If the personal goals or aspirations of each employee are important to the supervisor and safety is connected to those goals, employees will respond in kind. If it is not, they will also respond in kind.

When we genuinely show care, concern and interest in our employees and their safety they are more likely to buy from us. Why do we go back to the same restaurant, the car dealership or repair shop? It is likely not as much a result of the product or service, but how we are treated by the salesperson, wait staff or mechanic. Studies have found that we give our repeat business to the companies or people who treat us best.

To be successful we must show what sets safety apart from its competition. Safety competes against short cuts, complacency, "the way we've always done it," a culture that reinforces production as a priority. Safety must be positioned as a value, a higher value from its competitors, not as a priority against production or other facet of the operation. We must position safety in a way that helps the employee choose the best option by differentiating safety from its competition. David Hoffeld, in his book *The Science of Selling* calls it a "distinct value." I have adapted his rules for creating distinct value in safety.

Rule #1: Distinct value must matter to your front-line employees. Form a competitive advantage based on what matters to the buyer (employee).

For safety to effectively compete we must link the value that safety provides with what is important to the front-line employee. It is not likely employees care about regulations, but that is a mistake we often make. Selling safety based on government regulations is managing by edict and says, "We have to do it because the government says so or we can get in trouble." Distinct value is only persuasive if it is based on what the employee feels is important.

Rule #2: Distinct value must be unique. It cannot be replicated by the competition; otherwise, it is not unique. Safety as a product or service can help employees return home in good health. To enjoy their families, pets or hobbies, achieve personal goals or simply enjoy what they want most. Safety's competition cannot do that. Short cuts, complacency, even production or quality cannot do that. Not only can they not do that, they can prevent an employee from getting what they want most. Ironically, these same things negatively impact production and efficiency. "The more distinct value you can identify and communicate, the stronger your competitive advantage will be and the more likely buyers (front-line employees) will be convinced that your product or service (safety and health) is right for them" (Hoffeld, 2016, p. 71).

A distinct value statement might read something like this, "Safety is unique because of its moral significance. Safety creates a sense of community because when we create a safe work environment for ourselves, we also do it for our coworkers, contractors, our family and friends. A safe work and home environment prevents injuries to our immediate community, and prevents your family and friends from becoming a secondary victim. Short cuts and complacency only serve as obstacles to safety. How would you like to go home every day to enjoy what you want most?"

Another distinct value of safety is the intrinsic satisfaction we receive when we do something for others. Taking short cuts and being complacent not only goes against this advantage, it also puts others at higher risk of being injured. Defining distinct value can take many forms including statements, illustrations or pictures. Imagine a picture of your employee teaching his grandson how to fly fish (Figure 9.1). They are standing in a stream with their waders on and fly rod in hand, and his grandson is holding a beautiful rainbow trout. Below the picture, the caption reads "This is why I lock out my equipment." That is the distinct value of safety.

Throughout my career, I often attempted to appeal to an employee's fear of loss (injury) as a reason for buy-in. Fear of loss is a powerful motivator but must be accompanied by the means to prevent the loss. In other words, safety must be defined in operational terms as stated in Chapter 8. Fear of loss can also be framed as a loss of something important, for example, the ability to walk the dog, coach your son's basketball team or play a party parent in your daughter's Nutcracker performance.

A hurdle we face with buyers, especially those who have been doing the same thing for several months or even years is "Why change now?" "I've never been hurt." "I've been doing it like this since I started." We must be prepared to give the buyer a compelling reason to listen to us, to change if necessary; otherwise the status quo will remain the status quo. According to David Hoffeld (2016, pp. 54–55), in his book, *The Science of Selling*, the best way for the seller to answer is to help the buyer understand the problems that make change necessary. We must provide the knowledge they need to understand the problem and the implications. It is likely they face hazards that have never been brought to their attention or hazards associated with work done infrequently. We need to provide this technical knowledge, but equally as important is the knowledge they can provide us.

People, as with organizations, will not change unless they are dissatisfied with the current situation. The right knowledge of the hazards and the risk they pose must create a sense of dissatisfaction. They must feel the potential pain of, for example, not following

Figure 9.1 Making Safety Personal. Illustration by Masha Balac.

procedures or training techniques, or reporting a hazard. I am not referring to the pain of disciplinary action. That type of pain only ensures that minimum standards will be met, at best, and destroys trust. They must feel the negative impact of a serious injury to themselves, their coworkers and their families and friends.

The following steps are adapted from *The Science of Selling*:

Step 1: Find problems by conveying challenging insights and asking leading questions. For example, "How could you or a coworker get hurt performing this task?"

Step 2: Identify the cause and scope of the problem. "How bad could it be and who would it affect?" "Who are the potential second victims?"

Step 3: Ask deeper questions to help the buyers feel the painful outcomes of allowing those problems to continue. Ask "What are the most important things in your life?" and "How would a serious injury affect them?" (Hoffeld, 2016, pp. 53–57).

Asking these questions (even in a rhetorical fashion) helps to build trust, an important characteristic for any salesperson or successful leader. It also shows you care about them and their safety.

Are we asking the right questions and sharing knowledge? Or simply criticizing behavior? Unfortunately, criticizing behavior is often the chosen path because it is so easy to identify at-risk behaviors. Our challenge is to connect what they want most with safety.

PRAISE

Praise the slightest improvement and praise every improvement. Abilities wither under criticism; they blossom under encouragement.

– Dale Carnegie

This chapter is about knowing what the front-line employee wants most. Some wants are unique and some are universal. I talked previously about unique wants that are specific to the individual. Maybe it is family or friends, hobbies, sports, pets, worship activities or community service. Maybe it is more immediate. For example, the need for the right tool for the job. Universal wants are something we all want or seek. Praise is one of those universal wants. We all want to feel appreciated for our efforts. Recognition makes us feel like we are part of the team, that we are contributing to a goal, vision or greater good for the community. Psychologist Abraham Maslow, in his hierarchy of needs, calls praise and recognition "esteem" needs. According to Maslow's motivational theory, these needs motivate us to do the things we do. Most of us seek to belong, to interact with others, to have a feeling of accomplishment (Maslow, 1943). A recognition program based on the behaviors you want can help fulfill this need and ultimately help you get what you want, a worker who goes beyond minimum standards. It is important to note that this is short-term gratification. That is why it is important to be consistent with praise and recognition.

Research confirms, according to Joel Kurtzman (2010, p. 9) in his book *Common Purpose*, people working within organizations want to make a difference and be heard and recognized for their contributions or achievements. They want to be part of a winning team in a winning organization that has a mission and vision. The desire to be recognized is universal. Everyone wants to be significant in some way. I have never met anyone who came to work looking to get injured, and I have never met anyone who sought to be insignificant.

Defining and allocating rewards and punishments should be codified and communicated. Every group must know what its heroic and sinful behaviors are, and must achieve consensus on what is a reward and what is a punishment, according to Edgar Schein (2004, p. 127) in his book *Organizational Culture and Leadership*. Changing the reward and punishment system is one of the quickest and easiest ways to begin to change some elements of the culture. We often have walked into a "blaming culture" where the norm was to find someone to blame if something went wrong. That is how I managed safety as a supervisor. Who did it was more important than what happened.

Our reward and recognition programs often assume that more money is what most front-line employees want. Insufficient financial incentives can be a de-motivator, but more will not motivate an employee to go beyond minimum standards.

Identifying and commenting on something you genuinely like about the employee is also one way to build trust. It could be their work ethic, participation on the safety committee, how much they care for their PPE or how well they organize their work area.

Using praise increases your influence. By praising an employee for reporting a hazard, for example, you are labeling that employee as a safety-conscience person who cares about the safety of coworkers. According to David Hoffeld (2016) in his book *The Science of Selling*, positive labels are built on the notion that the behavior you publicly praise will increase. Rarely will anyone argue with a positive label or cringe after being praised. To the contrary, those praised will feel an instinctive need to live up to them. As a result, the next time you make a safety-related request, that person will think to themselves, "Of course I will follow through. I am a safety-conscience person who cares about the safety of my coworkers."

I frequently identified specific behaviors to target for praise. These were not random behaviors, but behaviors that were known to be contributing factors to injuries. We communicated the specific behaviors and praised employees when we observed the behavior. I used lapel pins inscribed with "Circle of Safety" when praising an employee, which turned into a conversation piece. Without exception, the employee immediately put the lapel pin on and I could sense their pride. When one employee saw another wearing the pin, it sparked a conversation that I imagined went something like this:

> Employee #1: "Hey where'd you get that pin? It's pretty cool."
> Employee #2: "The safety guy gave it to me."
> Employee #1: "How did you get one?"
> Employee #2: "He gave it to me for using the railing they asked us to use on the conveyors to prevent falls."

Praise and recognition are two of the least utilized and most effective tools in the safety toolbox. If you are not currently using these tools, now is a good time to start.

Here are some basic components of a successful recognition program:

1. Identify the specific behaviors you want. Select no more than 2–3 at a time. Once you have achieved habit strength, move on to other behaviors. Do not base your recognition solely on lagging measures such as completing a certain number of days without an injury. Behaviors should be defined in a positive nature. For example, completing training on time vs. failing to complete training by a specific date. Celebrating the achievement of a milestone based on a lagging measure is different and encouraged; however, it must be balanced with leading indicators, specifically the behaviors that bring about the achievements.

2. Communicate those behaviors so employees are clear on the expectations. Making the expectations clear ensures everyone knows what is important to you and what behavior is rewarded. It is also a key characteristic of successful safety cultures and successful leaders.

3. Provide immediate, specific, and positive feedback. Recognition should come as soon as possible after the behavior is noticed. Recognition that is delayed or vague losses it's impact. Rewards can always come later at a team meeting.

Many leaders and managers erroneously believe they cannot implement a recognition program because it is too expensive and not in the budget. The recognition itself can take many forms, most of which cost very little or nothing at all. Informal recognition such as a simple thank you for "following procedures when entering a confined space" or "staying current on electrical safety training" is appreciated. Handwritten thank-you notes are becoming scarce because it is just too easy to send a text message; but they are very personal and powerful. Neither costs a dime, but provides a significant return on minimal effort by reinforcing what is important and building goodwill among the recipients. In summary, recognition is low cost and high impact and increases your ability to influence.

SAFETY IS A POSITIVE

> *I am the supervisor, so it can't be my fault, so it must be your fault. Since this is your first offense, I have to put a letter in your file. On your next offense, I have to send you home for 1 day and put another letter in your file.*
>
> – Me, as a front-line supervisor explaining the disciplinary action policy to an employee for a safety violation

Dale Carnegie (1936) in his book *How to Win Friends and Influence People* said, "Any fool can criticize, condemn and complain – and most fools do." Nothing kills enthusiasm or ambitions like criticism or blame, yet too often we look for blame after an accident. We then view disciplinary action as a corrective action. By doing so, we turn a potentially positive environment into a negative environment. Or you might view it as compounding an already negative situation by focusing on the employee's behavior. We then compound the problem again by punishing the employee with disciplinary action.

Another universal want is to work in an environment that is positive. Where positive attitudes prevail, people work together and successes are celebrated. The work itself may not be fun, but working in a positive environment can make up for boring or difficult jobs.

Unfortunately, failures and mistakes are easier to spot than successes. Also, successes are expected. That is what everyone is paid for, right? Why praise success? Failures are a different story. We identify mistakes and believe we are correcting a problem. In reality, we are reinforcing safety as a negative and creating a negative environment. Working in a negative safety environment makes other negatives about the job more difficult to overlook. The flip side is that positive emotions and attitudes are contagious and resonate throughout your area of influence.

It is easy to take a negative approach to safety. Think about the negative connotations of the words we typically use when discussing safety:

accident, investigation, compliance, audit, disciplinary action, enforcement, mandatory training. How we talk about safety influences how employees feel about safety. Even safety professionals are not immune to thinking about certain elements of safety as negative and expressing them in negative terms. How often does the safety auditor show up at a site to conduct an audit? Are you glad to see him or her? Probably not. Maybe an audit should become a "benchmark" exercise or be seen as an employee engagement or training opportunity. I have even heard leaders with safety responsibilities talk about the boring nature of safety.

Kouzes and Posner, in their book *The Leadership Challenge*, mention that for leaders to foster a team spirit and to rally people around a vision, those leaders must practice positive communication. In other words, look on the bright side. "People actually remember downbeat comments far more often, in greater detail, and with more intensity than they do encouraging words. When negative remarks become a preoccupation, an employee's brain loses mental efficiency. This is all the more reason for leaders to be positive."

Kouzes and Posner are quick to point out that optimism does not mean being a Pollyanna, where nothing can go wrong. The authors quote research that finds persuasive leaders balance positive to negative statements by a factor of five to one. We should not project wild optimism while ignoring reality. Admit mistakes and address obstacles, but use more positive statements than negative ones. According to Posner, "It is that positivity that gets people to want to follow you and to believe enough that they are willing to do what you ask" (Kouzes and Posner, 2017).

According to Dan Petersen's *10 Basic Principles of Safety Management* (2001, p. 15, exhibit 2.3), "There is no one right way to achieve safety in an organization; however, for a safety system to be effective, it must meet certain criteria. The system must in part 'be perceived as positive.'"

Some people see safety as compliance, audits, inspections and investigations, all of which evoke thoughts of minimum standards at best and negativity at worst. Safety in this environment becomes an obligation. Safety is an opportunity to make a difference in the lives of those front-line employees we serve and to show leaders how safety can contribute to the business. The way we present safety affects how it is received and how people react to it.

Many employees have a preconceived negative notion about safety because of past experiences, just as I had for so many years. One of the core tenets in Dale Carnegie's (1936) book *How to Win Friends and Influence People* is that it is possible to change other people's behavior by changing one's behavior toward them. We must first have a positive attitude about safety. Be enthusiastic about safety and do not let anyone bring us down. If the supervisor has a positive attitude about safety, more likely the employee will too.

By translating features of a strong safety program into benefits we create a positive attitude about safety. For example, talk about incidents as an opportunity to identify a breakdown in the process and make an improvement. We do not follow lockout tagout procedures because it is a regulation, but because it helps to ensure everyone goes home with the same number of fingers they came to work with. Maybe "refresher" training should become "advanced" training.

I am not saying we should attempt to put a positive spin on a serious incident, but we should seek to understand why it happened and not look to manage through disciplinary action.

Everyone wants to be appreciated for their efforts and their contributions however small, as previously noted. Yet too often, we are quick to apply disciplinary action when an infraction occurs. We pull out that "read and sign" document we dutifully filed away three years ago when the employee was hired. It shows we did our job and the employee is at fault. I am not saying "read and signs" do not have any benefit or are not needed. The benefit comes in how we use them. Too often, we use them for disciplinary action purposes or to cover our own backsides.

In addition, disciplinary action is a poor accountability tool, as it looks back at what happened rather than forward by assigning corrective action. Using discipline in this manner tells other employees not to have an injury, but not always what to do to correct the breakdown. It places the focus on the employee's behavior rather than the management system where the breakdown likely occurred and perpetuates a negative atmosphere around safety.

To create a more positive atmosphere around safety, start by focusing the investigation on the management system or process that broke down and allowed the error to occur. Look for multiple causation rather than one simple act or condition. That is why I never use the term "root cause," because it insinuates that there is only one cause, "*the* root cause."

When an effort is made to identify all contributing factors, we may find other causes of operational errors, and we send a message that a breakdown in management systems may be at fault and not just the employee.

As a safety professional, I learned to effectively use this litmus test when pushed to consider disciplinary action for a safety infraction.

1. Is there a documented procedure that covers the situation? I don't mean "Yeah, we mentioned it in a shift briefing." I mean a documented, formal standard that is reviewed and updated on a routine basis.

2. Is there formal training associated with the procedure? Formal training can be classroom, computer-based training, or even a shift briefing, but key learning points and attendance should be documented.

3. Is the procedure effectively communicated? "Read and signs" do not count for much when they go back into the employee's file. Various communication methods should be used. Examples may include: posting on the company safety portal, newsletters, tailgate briefing, memos posted on bulletin boards. Each taken by itself has minimal impact at best, so look for multiple messages over a period of time. Multiple messages are even more critical for infrequent tasks.

4. Is the procedure or are key training points reinforced? Reinforcement comes through observations, inspections, audits and recognition. Yes, recognition. How many times have you recognized an individual or group for a specific behavior or procedure that was followed? If you recognized an individual, did you share it with the team so they know what behaviors are celebrated? For new procedures that require a behavior change, this is critical. Behaviors do not change over the weekend. It takes time and effort to reinforce the behavior to achieve behavior strength.

5. Did you measure compliance with the procedure? Can you say based on documented facts that compliance is at 50% or 60% or 75%?

By asking these questions, you are determining if expectations are clear, understood and reinforced. If you answered "No" to any of these questions, even one question, you have a management system issue that must be addressed. If you answered "Yes" to all questions then you have a personnel issue and the file should be turned over to your Human Resources representative or department head for action.

To date, I have not come across a situation where we could answer "Yes" to every question.

I firmly believe it is not safety's responsibility to be involved in disciplinary action, but it happens. When safety is seen as an accomplice to disciplinary action, we become an adversary to employees. Your motives will always be questioned and the term "investigation" will be linked to disciplinary action, further perpetuating a negative atmosphere around safety.

In summary, accountability is essential in all aspects of business, but particularly for safety. Unfortunately, accountability too often is synonymous with blame and negative consequences. To build a positive atmosphere, safety should never be involved with, associated with or linked to disciplinary action. We should know what they want most and disciplinary action is not on the list. The workplace is more enjoyable and the atmosphere more positive when we treat each other with respect and dignity.

Negative emotions do not lead to positive change. When we use disciplinary action for a safety infraction, we create a negative atmosphere. Everyone wants to work in a positive atmosphere. They are more engaged and are more likely to give that extra discretionary effort. Create that positive atmosphere by defining your safety program by its merits.

WHO ARE YOUR ALLIES?

We have over 40,000 employees worldwide at dozens of locations, I can't possibly write enough procedures, or conduct enough observations to reach everyone.

– Me, as a Junior Safety Analyst trying to figure out how to get one key message to every employee

I was a new Junior Safety Analyst responsible for influencing line operations managers and struggling mightily; that is when my director taught me the ABC rule, Allies Build Careers. Never underestimate the value of a good ally.

Many employees want to be involved to one extent or another. Find them and get them involved. These are your allies, your missionaries, the informal leaders and potential new formal leaders. These employees often are looked up to by other employees and as a result have the ability to influence others.

There are employees who want to be involved and aspire to move up into a supervisory position and eventually a management position, and some who want to contribute but are perfectly happy to stay where they are. You will also find a few who want no part of either. Focus on the previous two groups.

We want to feel a sense of belonging. To be a part of a community. Safety is the perfect platform to create this sense of belonging and sense of community. Many companies I have worked with aspire to have, and some have created, a work environment where employees at all levels look after each other's safety.

Let's be realistic. An eight-hour job may not be glamorous; it could simply be a paycheck to some. You can take that approach and get just that, eight hours of work. Or you can leverage an ally to help you get things done and, of course, sell safety. Yes, they can help you sell safety. Think back to Chapter 8 on communicating your vision and how powerful it could be for your ally to share your vision with coworkers.

We can take another page from the selling profession. Sales professionals often seek referrals from good customers because they know those referrals are likely to become new customers. If we apply the same logic, our allies, if properly cultivated, will talk about your vision to others, in effect referring you to other workers. An influential employee sharing your vision has a much greater impact than you sharing your own vision. Building allies makes your job influencing others easier.

Allies solve the numbers and logistics problem. Often, the safety department is made up of one person responsible for several hundred employees. Or a supervisor may be responsible for a team of 30 who are geographically scattered. I worked in the safety department of a company with over 200,000 employees worldwide. One person cannot possibly reach, observe or speak to every person, but a network of allies can! Provided with support, consistent messaging and direction, allies can share your message and model the appropriate behavior. Messaging designed for allies to share should be simple and direct. For example, people are generally risk-averse, but they must know the risk. Ask an ally to share pertinent facts and appropriate controls.

There is science here as well. Science that will help us sell. It is called "social proof," according to David Hoffeld (2016) in his book *The Science of Selling,* and it can help us conquer the negative perception of safety. When our allies are willing to model the behavior we want, it becomes OK for others. As more employees follow the behavior, it becomes the "safe" way to operate because that is what everyone else does, i.e., "If everyone else is doing it, I should too." Simply asking an influential ally to model a particular behavior may be enough social proof to start a behavior change among the team. Another way to use social proof is to ask your allies to share a success story, stories that demonstrate how a particular behavior or action prevented injuries or reduced the risk of a serious incident. When other teams or locations are having success (and being recognized), you want to have the same success.

Allies can be a great resource, but how do you find them? Sometimes they volunteer outright and sometimes they volunteer in subtle ways; for example, they may be the ones that always attend a safety meeting and sit on the front row. They may be the ones that make a suggestion or report a concern, or the ones who arrive to work early to ensure they are ready when the bell rings. They are the ones doing the job safely every day, day in and day out? Other times, they need to be asked to participate. Be on the lookout for the subtle hints or even better, just simply ask. They will get involved only to the extent they know safety is important. They will know safety is important if you have shared your vision with passion. Without question, to find your allies, you need to be on the floor. On the floor talking to them and sometimes working with them.

The next question becomes, how do you get them involved? The following suggestions are adapted from Dee Ann Turner's (2015, pp. 50–52) book *It's My Pleasure, The Impact of Extraordinary Talent and a Compelling Culture*:

1. Give them real responsibility. Activities that contribute to the vision and mission. Allies want to contribute in a meaningful way as soon as possible. Do you have a particularly tough fall protection issue? Put them in charge of a task force to solve the problem. Just be sure to give them adequate resources and training on problem-solving.
2. Ask them their opinion. They want to give their opinion and you may get it whether you ask or not, but they feel more respected when you ask. They usually have the answers before you. In addition, their trust in you will grow. Simply asking an employee for their opinion is also an informal form of recognition.

3. Create a clear development path. Help them craft a development plan that will position them for future career or personal growth. Involve them in cross-functional teams, safety committees, projects or a task force. Give them opportunities to interact with leadership. Providing an opportunity to present a topic at the safety meeting or during a toolbox talk helps build speaking skills. That is how I got my start in safety.

4. Advocate for your allies. They are often looking for sponsors. Provide public support and encouragement when they seek to solve a complex safety issue. Recognize your allies for their contribution to your injury prevention plan.

5. Allow your allies to fail without it being fatal. The best lessons are sometimes learned from our failures.

Allies can help implement change. Sometimes knowing what they do not want is just as important as knowing what they want. What they often do not want is change, the kind of change that comes with a new procedure or process that requires a behavior change. This fear of change must be overcome before you can move forward. Organizations sometimes change when they become dissatisfied with themselves, which often occurs after a serious incident. You may already have a pretty good safety performance, but you recognize a need to improve. Maybe your performance has plateaued and you want to break through to a higher level or fear performance may deteriorate. By sharing the reason for the change and the benefit to be derived with your allies, they become change champions for your cause and pave the way for the new direction.

In summary, the most successful leaders know they cannot be successful on their own. That is why they do not just monitor performance by reviewing the metrics; they monitor performance on the floor. They know the importance of having allies and look for them among the employees who do the job all day, every day. They invest in them and provide an opportunity to get involved and grow. Successful leaders view these employees as a potential starter on their team, a potential future leader or high impact employee worth investing in, not an asset to be managed. Who are your allies? Remember the ABC Rule.

Selling is about influencing behavior. Selling safety is about connecting what employees want most to safety. To be successful requires two types of knowledge. The first is the technical knowledge of the job that establishes our credibility and opens doors to conversations. The second is the knowledge of what the employee wants most and showing them how safety, i.e., following procedures and training techniques, can help them get it. Some wants are specific to the individual and others are common among most employees. Regardless, it takes good conversational skills and some creativity to make the connection, but once made, you begin to appeal to their wants and desires, and more importantly, you show them you care. The results are an employee group more willing to follow procedures and training techniques, report hazards and near-misses and espouse your vision for safety.

Three Things You Can Do Today:

1. Find out what one employee wants most and connect it to safety.
2. Show your appreciation by thanking an employee in writing. Remember soon, certain and positive.
3. Find one ally in your operation and share your vision, or simply give them an opportunity to get involved. Remember the ABC Rule.

REFERENCES

Carnegie, Dale (1936). *How to Win Friends and Influence People*, Simon & Schuster.

Hoffeld, David (2016). *The Science of Selling*, Penguin Random House.

Kouzes, James M. and Barry Z. Posner (2017). *The Leadership Challenge*, John Wiley & Sons.

Kurtzman, Joel (2010). *Common Purpose*, Jossey-Bass.

Maslow, Abraham (1943). A Theory of Human Motivation, *Psychological Review*, 50, 370–396.

Peterson, Dan (2001). *Safety Management: A Human Approach*, American Society of Safety Engineers.

Schein, Edgar (2004). *Organizational Culture and Leadership*, Jossey & Bass.

Turner, Dee Ann (2015). *It's My Pleasure, The Impact of Extraordinary Talent and a Compelling Culture*, Elevate.

10

LEAD WITH YOUR HEART

Last month we had 12 injuries on the day shift and...

– Me, running the monthly safety meeting
as a front-line supervisor

Safety, safety, safety, blah, blah, blah...

– What the employees heard

All supervisors had one additional assignment beyond their normal supervisory duties. My "extracurricular" assignment was to run the monthly safety meeting. I did not volunteer for it; it was assigned to me shortly after my promotion to supervisor. I still remember the exact moment when my boss hollered at me from across the airport breezeway, "Hey Karol, you are the new safety committee chair." My first thought was, "What does that mean?" I found out later it meant that I organized and facilitated the monthly safety meeting.

It was every other supervisor's responsibility to send someone from their team to attend and theoretically take back the information to share with their coworkers. I even took attendance and reported "no-shows" to my boss who followed up with the offending supervisor.

In preparation, I dutifully counted up the injuries from the previous month and compared year over year numbers. I reviewed injury reports for potential case studies or lessons learned that I could share. The meeting went something like this, I stand at a podium and in a somewhat monotone voice I read my findings to about 35 attendees never once looking up at my audience for fear of losing my place or maybe because I was afraid to see the boredom in their faces. Once I finished, I opened the floor up to questions and then dismissed everyone. My job was done and I returned to my work area and my normal duties. One thing I certainly did not do was connect emotionally with my audience. Even worse, they likely sensed my desire to get the meeting over with and get back to my work area. Success was judged by how many people attended. Not all leading indicators are created equal and this was a rather poor one.

If customers do not become emotionally connected to a product or service, they will not care enough to buy. The same is true of employees. That is why emotions are a determining factor in every sale. In this chapter, I discuss how leading with safety from the heart will improve your ability to sell safety to the front-line employee.

Safety can be a very emotionally charged topic, especially after a serious incident involving a coworker. Yet, we sell safety and attempt to influence change based on compliance (or lack of compliance) and corrective actions that resulted from investigations, audits, inspections or observations. Or as I at least attempted to, through injury statistics that do not mean much to front-line employees. We are engaging their minds with this approach, but not their hearts. We think in terms of preventing the next incident, which is not just our job as a supervisor or safety manager, but our obligation. Take the next step and view each incident as an opportunity to create a customer by appealing to their heart.

Set aside the moral aspect of safety for a moment and think of safety as a business. Businesses need customers to thrive. To build a customer base, businesses identify their target market or audience. Think of our employee group as the target market. To create a customer of our safety services, we need to put the employee first by providing the right tools and training. By that, I mean, providing tools and training that meet their needs and improve their ability to accomplish their job or task, not simply to ensure regulatory compliance. Refer back to the chapter on knowledge. This means getting them involved by asking them about their needs and what it takes to get the job done safely and more efficiently. To continue the business analogy, that means surveying your client base. Asking for their feedback shows your customers you care, that is leading with our hearts and that is when we start to build our customer base.

We can learn from Saint Teresa of Calcutta, better known simply as Mother Teresa. She was an Albanian-Indian Roman Catholic nun, missionary and Nobel Peace Prize winner. What did she know about leadership or finance, or marketing or safety management for that matter? Maybe nothing, but she did have a vision no more specific than to aid "the unwanted, the unloved, the uncared for." She was a different type of leader, a leader who led with her heart. She was humble, compassionate and had a passion to serve others. She may not have been a typical leader, but she left no doubt in anyone's mind what was important to her.

We knew she cared by her actions; they were undeniable. She connected with everyone in a very emotional way. She was optimistic, and an inspiration to millions. She could speak to and identify with the poor and sick, yet was admired by world leaders. Why? Maybe the better question is, how did she influence so many people? She was genuine, cheerful, enthusiastic, always positive, and she saw problems as gifts. She influenced people to contribute, and to work with her. Others joined her cause full-time. Mother Teresa was selling without intention to "sell." She served (and still does after her death) as a role model.

Our employees know what is important to us and whether we truly have their best interests at heart. Ask any front-line employee or an employee at any level what is most important to their boss. You will get an answer almost immediately because everyone knows by their actions. Do we lead with compliance or do we lead with our heart? Do we view our job, our assignments, our tasks as a service to front-line employees or as an enforcement action? Do we write procedures or conduct training simply to ensure compliance with regulations? Do we conduct inspections and observations to catch them doing something wrong and enforce the procedures? Do we conduct incident investigations to find someone to blame? That's how I managed safety as a front-line supervisor. As a result, my team knew what was important to me...compliance with procedures and disciplinary action for non-compliance.

Servant leadership is a philosophy in which the leader seeks to serve others, to put others before themselves. No one epitomizes servant leadership more than Mother Teresa. Do we view our job or profession as a service to our team or our employees? At her core, she was a champion of the poor, the "poorest of the poor." Let's be a champion of employee safety.

The following principles of a servant leader are adapted from Dee Ann Turner's book (2015, pp. 50–52), *It's My Pleasure: The Impact of Extraordinary Talent and a Compelling Culture*:

1. Do not expect others to do something you are unwilling to do. If you are conducting an observation, inspection or audit of a job or task you have never done, then spend a day at the job or shadowing an employee doing the job. The same applies if you

are developing or delivering training for a job you have never done.

2. Know your place in line. If the line is for lunch at a team meeting, be the last. If the line is for mandatory annual training, be the first.

3. Acknowledge that every team member is important. Never miss an opportunity to recognize their contributions. A handwritten note that is immediate, specific and positive has the biggest impact.

4. Share opportunities and privileges with those who might otherwise not have the opportunity. Provide advanced training that goes beyond the basic skills needed to do the job. For example, you might provide public speaking training or problem-solving training for members of the safety committee or task force.

5. Be inclusive. Focus on the needs of every team member. Build a community by providing teaming opportunities. Give team members an opportunity to start each shift or meeting with a personal safety moment.

Refer to the picture of Mother Teresa in Figure 10.1. What is the first thing you notice about her? Her smile. A real smile. A smile that radiates from her heart. People who smile manage, teach and sell more effectively. There is far more information in a smile than a frown. That is why encouragement is a much more effective teacher than disciplinary action. Encouragement is a form of coaching and recognition. Yet, disciplinary action is the first tool we often reach for as a corrective action. You might think it sits at the top of the hierarchy of controls.

Yet, as discussed earlier, safety is inherently negative, making it difficult for us to put on a smile. I believe if safety truly comes from your heart, you will smile. Smile, because you know you are working in their best interest, working to send more people home safe every day so they can enjoy the things in life they love the best. It works from a selfish standpoint as well. When you create a safe work environment for others, you are creating one for yourself.

I was attending a team dinner and having a safety-related discussion with a colleague from another location. We were engaged in a work safety-related discussion when another coworker, seated across the table, interrupted us to say, "Can we not talk about work after hours?" It caught both of us by surprise. We are safety professionals

Figure 10.1 Saint Teresa of Calcutta. Photograph by Michael Hoyt.

and we know why we do what we do. For us, safety was not work, it was personal. It was our passion. I seldom, if ever, get tired or bored discussing employee safety and certainly do not like missing opportunities to learn from others. Off-site conversations like that are often where creativity and innovation tend to peak. I also realize that when safety comes from the heart, you never miss an opportunity to talk about it, to learn and to improve. Needless to say, safety was not coming from this coworker's heart.

STATISTICS ARE NOT INHERENTLY HELPFUL

Safety is not about the numbers, it is about the people. Too often we get stuck in the numbers, as I did for so many years, and forget there is a person behind each number. The numbers are important. They are the scoreboard that provides proof that our program is either working or not. We use the data to help us make decisions. That is fact management, but too much focus on the numbers sends a different message. A message that says, "I only care about the numbers, not you," which can lead to managing the numbers rather than leading people.

We should provide the numbers in a scoreboard or graph format, but also understand that numbers alone will not influence behavior. If that was the case, we could simply post the information and be done.

What will effectively influence behavior is providing context behind the numbers and delivering the results with feelings and passion.

When we provide the context behind the numbers, people are better prepared to action the message or change a behavior. The context includes more than providing results of investigations or findings from audits. It includes the meaning of the numbers and the impact of those findings from both a personal and business perspective.

According to Edgar Schein (2004) in his book *Organizational Culture and Leadership,* "One of the most important dimensions of culture is the nature of how reality, truth and data are defined. What is a fact, what is information and what is truth depends not only on shared knowledge of formal language, but also on context." Collecting and distributing data is easy. That is what I did as a supervisor and continued to do in my professional career. I just assumed everyone could interpret the data and take appropriate action. Not only was I wrong, but that is not what a front-line employee wants or needs. I have seen many bulletin boards full of beautiful charts and graphs, and I suppose some front-line employees have not only an interest, but can interpret the data. What most employees want and need is useful information, much like senior leaders; "Don't give me data without interpreting the data." More specifically, front-line workers want to know what they need to do to avoid becoming a data point.

I was working as a corporate safety manager when I was assigned to support a particular executive leader. The division employed approximately 2,000 employees at several locations across the United States. The employees were mostly managers and supervisors. Our first meeting went something like this:

Me: "Good morning "Steve." I'm Pat Karol and I'm here to support your safety efforts."

Steve: "Thanks Pat, but we just reduced our injury rate below 10.0 for the first time and we won't be needing your help. Besides, any safety work that needs to be done is handled by my Human Resources (HR) team."

Me: "Human Resources?"

Steve: "Yes, I took safety away from the front-line managers to free up their time to do more important stuff." (I paraphrased here)

Me: (to myself) OK, I am not going to influence change today. I need to change tactics real quick. I need to leave and live to fight another day.

Me: (out loud) "Steve, thanks for your time. I appreciate your candor and can appreciate the challenges you face. I'll be available if your HR team needs support. By the way, do you mind if I meet with your HR Director?"

Steve: "Sure, go ahead."

Me: "Thanks, have a great day."

He was proud of his accomplishment, although he could not point at any activity that impacted the rate, for better or worse. I followed up and started meeting with the HR director on a routine basis. I employed the ABC Rule "Allies Build Careers" (and they help you get stuff done) from Chapter 8. I was eventually able to use her influence with the executive leader to get 15 minutes at the next quarterly operations team meeting. Approximately 50 operations managers and support staff attended to discuss the previous quarter's performance and strategy going forward.

I did not prepare a speech or an elaborate PowerPoint deck. I simply wrote the description of the previous five injuries on a Post-it note and placed the note under the notepad at every 10th seat. When it came to my appointed time on the agenda, I asked everyone to look under their notepads and if you find a Post-it note please stand up and one by one read the information. It went something like this:

First Manager: "Employee slipped on food spilled on floor, fell and suffered a broken elbow, $14,000."

Second Manager: "Employee was lifting a 22.7 kg (50 lb) bag of flour from a pallet on the floor and suffered a back injury $18,000."

It went on until all five managers read from their Post-it notes. When they finished, I simply said, "That is the equivalent of a 10.00 Total Recordable Injury Rate. 1 in 10 will be injured seriously enough to require medical attention beyond first aid. That is where you are today. If you are OK with that, I'll move on. If not, I am here and can help. My job is to make you the hero. To reduce costs associated with worker's compensation, reduce turnover (a big problem for this leader) and reduce operating costs related to overtime pay to cover injured employees." I provided context to the rate and personalized the numbers. In addition, I provided at least a partial solution to a big problem and eventually established a very good working relationship with the leader. His team's efforts resulted in a substantial reduction in the injury rate over the next two years.

Another option to add context is using a benchmark or comparison. For example, I could have compared his rate of 10.0 against the General Industry average published by the Bureau of Labor Statistics, which was 3.8 at the time, or compared his rate against the other operating divisions. I chose to keep those figures in my back pocket for future use. As their rate improved, I did eventually leverage their competitive spirit.

Context in the form of pictures and stories or personal notes can have a powerful influence as well. Personal notes in the form of a commitment are particularly powerful. For example, post personal pictures with a personal commitment that is signed by the employee. That's context, that's personal and that's powerful.

I later supported another executive leader for a facilities operation, which covered warehousing and logistics, building maintenance, sanitation and janitorial work. When I arrived, the injury rate was above 7.0. This leader was a strong sponsor and believed that injury and illness prevention was not only the ethical thing to do, but it resulted in improved efficiency and, depending on the client, could be a market differentiator. I corroborated with his team, and we built a strategy with defined goals and held the team accountable. I provided monthly updates on performance and feedback from the field. We met quarterly with his line operators to review progress toward the goal and make adjustments as needed. As we started to improve, I dutifully reported the injury reductions. Fewer people getting hurt is a good metric to track, but it seemed to focus too much on the negative angle. Eventually, I started calculating how many more people were going home safe year over year. It was more of an approximation than an exact science since headcount routinely changed, but the message was positive, heartfelt and provided more context than simply injury reduction numbers.

Statistics are not inherently helpful. What is important is the context and the meaning we provide behind the numbers. The context we provide personalizes the numbers and helps employees and their supervisors understand why they are important.

HAVE A "MAKE SAFETY PERSONAL" CONVERSATION

Let's tie it all together in a conversation with the "Make Safety Personal" conversation framework.

I was introduced to a communication tool by Karen Warner, Chief Executive and Managing Partner of Tangible Group. It is actually a format for having a strategic conversation with an executive leader. I have used this format successfully numerous times, and it helped me organize my thoughts and put them in a succinct conversational format.

We can have a similar conversation with employees, so I adapted Warner's tool and call it the "Make Safety Personal" conversation, which provides a framework for communicating and verbalizing safety from the heart.

It works like this:

1. *State your vision.* Start by stating your vision. "My vision is [fill in the blank], or I believe [fill in the blank]." This is your opportunity to lay the foundation for a shared vision. If you don't have a vision, that is OK! Start by explaining why you are a safety professional or supervisor and why you believe safety is important. The way we present our vision or safety in general will affect how people respond. If your vision is presented as if it were an edict to be followed, you will be lucky to have any followers. If your vision is based on compliance, "My vision is to be in 100% compliance with company and government regulations," you will be lucky to achieve a minimum standard. When we are knowledgeable and passionate and convey positive emotions about our vision or why we believe safety is important, those emotions are conveyed to the audience and improve people's ability to understand and follow your lead.

There is an important distinction to be made between sharing your vision and explaining why. According to Simon Sinek (2009) in his book *Start With Why, How Great Leaders Inspire Everyone to Take Action*, a vision is aspirational, something you want to achieve. Your "why" is a statement of fact. It is a statement that describes a current condition. For example, I am a safety professional because, as a supervisor, I saw firsthand how a serious incident cost a worker his leg and affected him, his family and coworkers. I want every worker to go home safe. That is "why" I do what I do. I believe that "every injury is preventable." That's aspirational. That is my vision that I am working toward. You can use either to start your "Make Safety Personal" conversation.

2. *Explain why your vision (or why statement) matters.* This is your opportunity to make safety personal, to connect safety to what people want most. Ideally, you would know something

personal about someone in your audience or the person you are speaking with. If it is a large audience that you do not know, then use a personal example or story. I have used my personal example as a Little League Umpire and tied it to the use of personal protective equipment (PPE). I have also used my story as a supervisor and how a serious injury to a team member affected me. Explain why it must be done this way. "Joe, I know you work with power tools in our factory and enjoy fishing as a hobby. Can you imagine how a serious hand injury would affect your job performance and ability to enjoy your hobby?" That is why we stress the importance of adjusting machine guards and wear eye protection. Fear of loss and desire for gain are both effective motivators, so do not be afraid to use either or both. This is also a great time to tell a success story that demonstrates why your vision is important. Your success story should highlight how a similar group or peer group benefited from embracing your vision.

3. *Define your vision (or your why statement) in operational terms.* It is not enough to say "be safe," because it means something different to everyone. It must be defined in operational or actionable terms as discussed in Chapter 8. Define what will have to happen at work to achieve your vision or to be consistent with your "why" statement. An actionable term may be reporting and investigating every incident no matter how slight. Another example is conducting a pre-task analysis before every job that is done infrequently.

4. *Invite dialogue by asking questions.* This is a critical step in achieving support and making your audience more receptive to your vision. In addition, everyone wants to know their opinion is valued. Attempt to see things from their point of view. Besides engaging your audience, the right questions can have a powerful and positive effect on their behavior. Your questions should prompt them to share their opinion. What is your perspective? Do you have a better idea (if post-incident, what could we have done differently?) If you could change one thing about your safety training or [fill in the blank], what would it be? What are the roadblocks? Above all, listen carefully. Answers to these questions provide insight into the beliefs that shape their current behaviors whether they are at-risk or behaving safely. Listen for concerns and suggestions for improvement,

but most of all listen carefully for passion in their voices and you might just find a new ally.

5. *Clearly state an action plan.* Clarify what you will do and what the employee will do, and the timeline for each. When roles and expectations are not clear, confusion will result and the culture will suffer. Once the action plan is agreed upon, follow up frequently. Failure to do so will result in a hit to your integrity and your ability to sell will diminish. Express a sense of urgency.

Throughout your "Make Safety Personal" conversation, convey your beliefs with profound confidence. Show enthusiasm for the possibilities that can be achieved. People learn to the extent they are motivated. You provide the motivation to learn by being knowledgeable, passionate and conveying positive emotions. Those emotions are contagious and improve your audience's capacity to understand and retain information.

Why have a "Make Safety Personal" conversation when we can simply initiate change by mandate or edict passed down from above? You can see evidence of this in statements such as "safety is a condition of employment." On the other hand, the best leaders facilitate change by engaging their stakeholders in the process and setting expectations rather than issuing mandates and edicts. When employees realize what is expected of them, but perceive some personal control in setting goals and determining how to reach them, they are more likely to own the process. The "Make Safety Personal" conversation allows you to begin the process of setting expectations and engaging your employees in the process without mandating policy and establishing edicts.

A "Make Safety Personal" conversation might go something like this:

Example #1

1. You (state your vision): I believe every injury and illness is preventable or rephrase it as a question. Do you believe every injury and illness is preventable?

2. You (state why your vision matters): Preventing every injury and illness is important because I have seen how one serious injury can impact many people. I believe every injury and illness is preventable because we know the contributing factors of every incident, we have the knowledge and have the capability, the training and procedures and can implement the systems that

will prevent injuries and ensure we will all go home today. That should be important to all of us. Everyone here has family, friends, hobbies, pets they want to enjoy. Vacations to take, movies to see or Moe, in your case, trout to catch. In my case, I have Little League baseball games to umpire. That's why I wear PPE on the job.

3. You (define your vision in operational terms): Believing that every injury is preventable means instead of saying "be safe today," we acknowledge hazards and what behaviors are required to avoid injury. More specifically, we have a large shipment of material arriving today and there will be higher than normal forklift activity in the warehouse. If you need to be in the warehouse, stay in the pedestrian lanes, stop and look at intersections. If you encounter a forklift, make eye contact with the operator, stay at least 3 meters (10 feet) away and communicate clearance before crossing an intersection. If you are a forklift operator, yield to all pedestrians, slow down at blind corners and stay within the speed limit. These are the behaviors that must happen to prevent the next injury. Your safety is a worthwhile investment of our efforts.

4. You (invite dialogue): "I am interested in your opinion. If you could change one thing about our safety program that helps us prevent the next injury and achieve our vision, what would it be?"

Employee #1: "Well, the training is boring and I don't get anything out of it."

Employee #2: "And some of our procedures don't make any sense! Why do I have to wear eye protection in the warehouse?"

You (invite more dialogue): "If we invest in new or revised training, review our eye protection procedures, how will that help us achieve our vision?"

Employee #3: "It has been the same training for the past five years. It doesn't apply to us. We have new equipment and other changes that we need to know about."

You also might hear critics or crickets. That is OK. Ask them to ponder the question. They might need more information. They may need to discuss what makes the training boring and what can be done to relate it to the job. You can facilitate this discussion in a follow-up session or informal one-on-one conversation. Just do not wait to continue the discussion and be sure to share results with all employees.

5. Agree on an action plan: "Your feedback is valuable. I feel like we are a step closer to preventing the next injury. Would you be willing to join me on a task force to review the training with our vision in mind? As a follow-up, I will be on the floor for the next two weeks to speak with each of you about what we can do together to achieve our vision."

Example #2

1. You (state your vision): "I believe safety is the ultimate team sport because we can only achieve our goal of zero injuries by working together, much like a sports team striving to win a championship."

2. You (state why your vision matters): "Approaching safety as a team sport matters because when we commit as a team we can make progress toward our vision. When we all play our position, we create a safer environment for everyone. Preventing injuries and illnesses takes more than a manual full of procedures and training. It takes everyone making a personal commitment, to view each other as a teammate whether we work alone or in a group. We all have a personal stake in the game. By taking a team approach to safety, we ensure that we all go home today to enjoy family. Dave, for you, I know that means being able to coach your daughter's soccer team."

3. You (define your vision in operational terms): Working together means we identify even the smallest hazards, communicate, make the right calls and execute the right plays. Working together as a team to prevent injuries means, for example, when we have changes in our work environment that introduces a new hazard, we communicate that hazard to each other and take action to eliminate or reduce the risk.

4. You (invite dialogue): "I am interested in your thoughts. How can teamwork help us create a safe work environment?"

Employee #1: "I am a forklift operator and I almost ran into a guy walking outside the pedestrian lane the other day because pallets of stock created a blind corner and he was not paying attention. If we are all looking out for each other like teammates, that wouldn't happen."

Employee #2: "I operate a stock picker in the warehouse. I almost ran into a forklift operator last week. If I know when they plan more than normal forklift activity, I can adjust my schedule and work area."

You: "What one thing could we start doing that would help us prevent an incident with a forklift?"

Employee #1: "We could post color-coded placards to indicate heavy, medium or light forklift operation days and place the placards in areas where the activity will be the heaviest."

You: "I hear you say increase communication between the forklift operators and others who work in the warehouse. Is that right?"

Employee #1: "Yes, that will work."

5. You (agree on an action plan): "Strong and clear communication between groups is a key characteristic of successful teams. I'll put some ideas together and bring this group back together next Monday for your feedback. We will decide on the next steps at that time. Thank you for your input. Together, we can build a strong team that will benefit all players."

You may not have won everyone over yet, but you opened the skeptical employees up to your vision.

Leading with your heart is knowing and understanding why we do what we do and conveying that with passion and enthusiasm. It means you are thinking of others before yourself. Using your heart to engage their heart.

Three Things You Can Do Today:

1. Have a "Make Safety Personal" conversation with one person. Practice honing your message with an ally first, but don't wait. Download the Make Safety Personal Conversation Worksheet at www.karolsafety.com

2. Post photos of things that are important. I like to use the example of Moe who likes to fly fish on weekends with his son with the caption "This is why I lock out my equipment." That image connects safety to something very personal not just for Moe, but for anyone who knows Moe.

3. Start every meeting with a personal safety moment or "safety share." Everyone has experiences they can share. Maybe a near-miss or an incident they observed or lessons learned. Use this opportunity to tie the safety moment to your vision and do not forget to publicly and formally recognize the person for sharing.

REFERENCES

Schein, Edgar (2004). *Organizational Culture and Leadership*, Jossey & Bass.

Sinek, Simon (2009). *Start with Why: How Great Leaders Inspire Everyone to Take Action*, The Penguin Group.

Turner, Dee Ann (2015). *It's My Pleasure: The Impact of Extraordinary Talent and a Compelling Culture*, Elevate.

SECTION **3**

WRAP IT UP

11

HURDLES WE FACE

As with any endeavor that involves change, you will face hurdles big and small. I have listed the most common hurdles that I have faced. If you know they are coming and are prepared, your chances of succeeding will improve.

1. The 80/20 Rule (some we will never convince). Vilfredo Pareto was an Italian economist and mathematician at the turn of the 19th century. He determined that 20% of the people owned 80% of the land. His studies eventually became known as the Pareto Principle. We know it as the "80/20" rule. In business, as in the safety field, we use the Pareto Principle or analysis to identify areas to focus resources to obtain the greatest impact. Based on this principle, we can assume that somewhere around 20% of the front line will never sign on to your vision. Accept it and focus on the 80% that do.

2. Company culture is the elephant in the room. The safety culture is impacted by the organization's culture. A quote that is been attributed to more than one management guru, but most frequently to Peter Drucker, goes something like this, "Culture eats strategy for breakfast." In other words the best strategy will come up short if the culture does not support it. Cultures come from the top of the organization and are based on the beliefs, values and assumptions of leaders and their experiences. More informally, culture has been defined as "how we do things around here." For example, on-time performance dominated and defined our culture when I worked in the airline industry. It was the key metric that everyone knew and worked toward achieving. Leaders clearly valued on-time performance. It was not an inherently bad culture, but it impacted the safety culture. We learned to leverage the corporate culture by showing how safety impacted on-time performance.

If you are struggling with leveraging the corporate culture, my suggestion is to focus on the area where you have the most influence. Maybe it is a particular department or site. Maybe it is a shop where you have a strong ally. Maybe it is a person with influence and a broad span of control. Look for opportunities to start the discussion. For example, a near-miss that had the potential to be serious or a trend within your industry. A benchmarking activity with another organization can also provide the inspiration to change or at least start the discussion. We can create learning experiences based on the results of these activities that support a strong safety culture.

3. New hires and high turnover rates. High turnover rates mean lots of new-hire employees. That can make getting your message out an ongoing chore and uphill battle. Imbed your message in new-hire orientation. Use various communication means including video and stories. Define your vision in operational terms and form a 10-point "Make Safety Personal" pledge that each new hire signs and posts in their work area. If you hire groups of people at a time, you might opt for a poster-sized version they all sign. You might even make it a celebration event at the end of orientation. Be sure to post the commitments in a common area. Visible displays of commitment can be very powerful influencers. Pair your allies up with a new hire as a "buddy" for the probation period. Provide your allies with the "Make Safety Personal" pledge and have them review and reinforce each line item with the new employee.

4. Number of employees and geography. Getting one message out to hundreds or thousands of employees in one location is a challenge, but when those employees are spread across the country or globe, the challenge is compounded. This is where your allies in the field show their value. At one company I worked with, we established and grew a network of allies (we called them Safety Leaders). We provided structure in terms of a job description, training through monthly conference calls and an annual Safety Leader conference to align efforts.

Ensure your allies are clear on your vision and expectations for communicating your vision, what it means and their role in achieving progress toward the vision. Organize periodic conference calls to reinforce your message. I held monthly conference calls with the site operations managers, many of whom were also my safety allies. They could invite anyone on their team to join. The purpose of the call was to share a success story. It could be the lessons learned from a near-miss incident and the action they took, or a hazard they identified and the engineering controls they put in place, or the success of a safety committee and their level of employee engagement. I prearranged the call by identifying a success story and helped the site manager prepare the presentation. The executive leader and middle managers were on the call and publicly recognized the manager.

Over the four to five years I facilitated these calls, I never had a site manager that was not excited about talking about their success and proud of their accomplishment. I heard from many site managers about how their pride was reflected in each team member. Within a few days of the call, I published the presentation via email, which gave the executive leader an opportunity to once again recognize the site manager, which in turn created a learning experience that reinforced the vision and safety culture.

5. Trust. I talked a lot about the importance of building trust in earlier chapters. Trust works both ways. If the front-line does not trust management to support them if they make a decision based on risk, they will quickly lose faith or confidence in any safety initiative. In addition, if management does not follow up on reported safety concerns or hazards in a timely manner, your job as a safety manager or supervisor just got tougher. I use the term "management" collectively to mean supervisors to middle management up to executive leaders. If you are faced with this situation, focus on what you can control, where you have influence or who you have influence with. Encourage prompt and thorough follow-up on all action items.

6. Management styles. Management styles vary from supervisor to supervisor, from manager to manager and from leader to leader. Some, more specifically the command and control (just tell them what to do and enforce it through disciplinary action) or theory X manager (employees are generally lazy and you need to entice them with carrots to get them to work) will view the approach I take as a soft, "country club" approach. There are situations where a command and control style or transactional style of management is most appropriate. Further, in over 40 years in the workforce, I have seen firsthand how establishing a vision, engaging employees and recognizing their efforts

can far exceed the minimum standards that you get from other management styles. Stay your course and focus on incremental changes, quick wins and even small wins. Successful change does not come from a keenly written safety policy. It comes as the result of persistence and building on a series of incremental successes that are celebrated. Soon they will be coming to you to ask "What is your secret?" You can smile and say "It is all about making safety personal."

CONCLUSION

It started with anonymous feedback. A question from a session I gave at the Georgia Safety, Health and Environmental Conference in Savannah five years ago. "How do I get my employees to do what I want them to do?" That simple question from someone I will never know, prompted me to reflect on my work career and evaluate my activities, my successes and failures. One thought I came back to more than a few times was, if I was starting over today as an operations supervisor or safety professional, what would I do differently?

I learned that to build an extraordinary safety culture that has a positive impact on an employee's life (and by default the organization), you must understand two things. First, there is no silver bullet, shortcut or quick fix. Extraordinary safety cultures are not built overnight, even when starting from scratch, but especially when attempting to turn a distressed culture around. In both cases, it takes constant, steady pressure. Steadfast leadership at all levels. Persistence and patience. The ability to get back up quicker than you fall, every time. Most importantly, it takes purpose and passion. It takes knowing why you do what you do and having a passion for it.

My purpose and passion is to serve front-line employees because they deserve to work in an environment where they can do what they do best without fear of being injured. I do that by writing and speaking to operators, safety managers and human resource professionals about making safety personal. I do it because at heart I am an operations guy and I have an affinity for the folks who work on the front line for a living. I have seen how the making safety personal components of vision, knowledge and heart can have an impact.

Second, you must be satisfied with the fact that you may be one of the few or even the only one to see the impact. There is no instant gratification for your efforts, because when we get it right, nothing happens. Recognition, if it comes, will not likely be immediate, specific and positive. Our employees will come to work and go home the same way every time. When will you know you are successful and have built an extraordinary safety culture? When your employees come back to work after the weekend and tell you about the fish they caught, then don their PPE. You will know. When they tell you how they coached their daughter's softball team in an upset of the first-place team, just before conducting the safety briefing (without using the word "safety"). You will know. When they tell you their plans for visiting their new grandson, just before they begin the pre-task analysis. You will know. You will know you have built an extraordinary safety culture and you did it by making safety personal.

By sharing my story, experience and lessons learned, my hope is that you can shorten your learning curve and start to have an impact today. Maybe you are thinking, "I am just one small cog in the wheel. Besides, no one cares around here." My suggestion is to start somewhere. Start small. Gain one ally at a time, win over one employee at a time, because you cannot do it alone. There is too much at stake not to give your best effort. Maybe you thought to yourself, "Yeah! That makes sense! I can start today and have an impact! I will influence change. I will be a compelling example of 'making safety personal!'" I say, share your passion with someone today!

If I can tell you to do just two things, the first would be to build trust with your front-line employees. The more employees trust you, the more receptive they will be to your vision and your message. Trust makes the decision to be safe easier. Second, lead with your heart. Communicate with emotion, enthusiasm and passion. Show excitement for your safety program and its benefits to your employees. Celebrate every success, big and small. When you do those two things, they will know what is important to you and will respond accordingly and with consistent behaviors. Do not underestimate your ability, impact and influence as a seller. You have a safety brand whether it is intentional or not, so be the best dressed, best spoken, most knowledgeable safety salesperson you can be.

If you do not have a vision or it is not yet clearly defined, do not worry, it will come. If you are looking for a starting point, here it is, "Every injury is preventable." Believing every injury is preventable is the equivalent of a growth mindset. It is more than a stretch goal. You might even call it an impossible goal. That is OK! We only know what we are capable of by setting impossible goals.

I have included a Safety Leadership Self-Assessment as an addendum in my book. It is based on the lessons I have shared. I encourage you to complete the self-assessment and build your own Personal

Action Plan. You might even consider sharing it with your manager or colleague and develop an accountability relationship. To download a pdf version, go to my website at www.karolsafety.com. My vision for this book, as for the talk that preceded this book, is *To be a compelling force behind which you will feel inspired to Make Safety Personal.* I hope today I am one step closer to achieving my vision.

One last thought, actually a request. I want to hear from you! Share your experiences and lessons learned. Have you experienced similar challenges? How did you overcome them? Go to my website www.karolsafety.com to share and I will post your lessons learned and maybe you will shorten someone's learning curve and ultimately send more employees home safe by *making safety personal.*

Instructions:

1. Select three to four of the "Needs Improvement" questions from the Leadership Behavior list. This could be an activity you will start doing, improve upon or stop doing.
2. Identify a target frequency for completion of the activities.
3. Identify the action you will take to address the item you identified as "Needs Improvement."
4. Share these actions with your local team and your direct manager and ask for their support.
5. Review progress to this action plan each month to make sure you have incorporated these actions into your "pattern of management."

Make Safety Personal
Safety Leadership Self-Assessment

Name_____

Date_____

Leadership Behavior	Good	Needs Improvement	Comments
Vision: Finding and Communicating Your Vision			
Do you have a personal vision for what safety means to you?			

Do you define your vision in operational terms? (not using the word "safety")			
Do you emphasize cooperation and commitment over compliance when talking about your vision?			
Do you use stories to reinforce your vision?			
Do you express your vision often and use multiple means of communication?			
Do you communicate a personal commitment to your vision? Handwritten?			
Do you share your vision with vendors, customers and communities?			
Knowledge: Recognition, Safety as a Positive, ABC Rule			
Do you provide recognition that is immediate, specific and positive to employees for behaviors that support your vision?			
Do you celebrate team successes based on strong leading indicators?			
Do you recognize employees who do their job safely day in and day out?			
Do you use personalized handwritten notes to recognize individual performance?			
Do you translate features of your safety program into benefits? (not compliance)			

Do you provide training or other opportunities for employees to get involved beyond regulatory requirements?			
Do you use "allies" to help you get your message out and get things done?			
Heart: Passion, Context, Make Safety Personal Conversation			
Do you provide context behind injury rates and numbers?			
Does your voice and body language reflect passion for employee safety?			
Do you set the example by being the first to complete training, follow procedures, etc.?			
Do you provide frequent opportunities for employees to provide input or report concerns?			
Do you ask individuals questions about what they are doing to proactively prevent injuries or support your vision?			
Do you follow up on reported concerns, investigation and observation results in a timely manner?			
Do you ensure that your employees have time to participate in injury prevention activities? Do you participate?			
Do you set clear expectations that support your vision?			

Do you conduct Make Safety Personal conversations on a routine basis?			
Other			

Make Safety Personal
Personal Action Plan

EXAMPLE ONLY			
Name:		**Completed**	
Frequency	**Action**	**Yes**	**No**
Weekly	Conduct at least one Make Safety Personal conversation		
Weekly	Identify and thank an employee in writing for supporting the vision		
Monthly	Review the outcome of a safety committee meeting with employees		
Each Incident	Personally follow up with the injured person to express concern for their health and ensure corrective actions are put in place to eliminate future occurrences		

Safety Personal Action Plan			
Name:		**Completed**	
Frequency	**Action**	**Yes**	**No**